THE STAR BOOK

The Black Eye
Galaxy (M64) in
Coma Berenices is a
prominent member
of the Coma-Virgo
cluster of galaxies.

THE STAR BOOK

HOW TO UNDERSTAND ASTRONOMY

PETER GREGO

D&C
David and Charles

CONTENTS ....

The beautiful Horsehead Nebula in Orion is very challenging to view through even large amateur telescopes, but it is captured here on a CCD image taken through a 127mm refractor (filters used).

Astronomical CCD image of the Great Spiral in Andromeda (plus its companion galaxies M32 and M110). Field of view is two degrees (four times the apparent width of the Moon).

FOREWORD

BY SIR ARNOLD WOLFENDALE FRS, 14TH ASTRONOMER ROYAL.

14th Astronomer Royal

It is a pleasure to write this brief foreword to Peter Grego's latest book. Peter is a very well-known popularizer of astronomy; he can be trusted to make his topics both fascinating and accurate, and this book is no exception.

The potential market is enormous in that it caters for those without instruments of any form, as well as those lucky enough to have binoculars or a telescope. Thus the increasing number of retirees, as well as youngsters, will find this book useful and fascinating. And what a feast of fascination there is. Constellations are dealt with on a quarterly basis and those living in the Southern hemisphere are not neglected. The host bidding his visitors farewell on a dark and clear night will be able to show off mightily, indeed recruits to the astronomical cause will surely flow.

A highlight of the book is an in-depth study of planets, comets and bright asteroids. If you can see it in the sky then there's something in this book to tell you more.

Finally, and unusually, there is a section on light pollution.

Read on and enjoy...

Sir Arnold Wolfendale FRS
14th Astronomer Royal.

Earthshine illuminates the dark side of the young crescent Moon, imaged with an undriven digital SLR.

The Bubble Nebula
in Cassiopeia,
imaged using an
80mm refractor and
astronomical CCD
camera (filters used).

Introduction

People have gazed at the starry skies with wonder and awe ever since the first sparks of human consciousness flickered in the minds of our distant ancestors. Astronomy is the oldest of all the physical sciences; records of celestial events go back many thousands of years to the beginnings of the earliest cultures. Celestial cycles – the movement of the Sun, the phases of the Moon and the annual seasons with their grand parade of stars and constellations – gave humans a practical means of timekeeping; later, the great seafaring civilizations of antiquity learned to navigate by the stars. Things have come full circle; the most accurate clocks we know of are to be found in the regular flashes from exotic stars called 'pulsars', while above us in orbit GPS satellites enable navigators to pinpoint their position on Earth to within metres.

Examples of three celestial objects that can be enjoyed with the unaided eye, binoculars or a telescope: the Orion Nebula (left), a stellar birthplace; Mars (centre); and the Pleiades (right), a cluster of young hot stars.

There's a great deal to see in the night sky. Whether you use binoculars or a telescope, or even if you have no optical aid at all, there are enough celestial sights to keep anyone enthralled for a lifetime. Stars and constellations are permanent fixtures; others, like the planets, move against the celestial backdrop and appear to change over time. A few phenomena, such as meteors and eclipses, are fleeting but spectacular.

So much can be seen in the night skies without optical aid. It's fascinating to spend time learning the layout of the skies, the positions of the main constellations and the names of the brightest stars – essential, even, if you would like to take your enjoyment of astronomy to a higher level. There are advantages to living in an urban area, where sky-glow caused by city lights drowns out all but the brighter stars. Although just a few hundred stars might be seen with the unaided eye from a dark enclave in a city centre, the night sky appears less crowded and the patterns of the main constellations are easier to trace. Under a dark rural sky where several thousand stars can be seen, the heavens can appear so congested with stars that even experienced astronomers can become a little disorientated.

Using binoculars

Eminently portable, with wide fields of view and low magnification, binoculars give unlimited freedom to roam the night skies – they are the ideal starter instruments for anyone who is new to stargazing. Binoculars gather more light than the eyes alone and reveal many of the night sky's hidden treasures; they also give the startling (although illusory) impression of three dimensions in space. The colours of stars are especially noticeable through binoculars, and many star clusters, nebulae and galaxies, as well as countless glorious starfields, can be viewed with their aid.

Comparisons of views through different binoculars: the Moon, as seen through 7x30s (left), 12x50s (centre) and 25x100s (right).

The power of binoculars is identified by two figures – one denoting their magnification and the other the size of their objective lenses. For example, a 7x30 binocular – the smallest practically useful for stargazing – magnifies seven times and has 30mm lenses. Even 7x30s can show several hundred thousand stars and numerous deep-sky objects, while more than a million stars and thousands of deep-sky wonders can be seen through a large 20x80 binocular.

Using a telescope

With their greater light gathering ability, telescopes deliver detailed, magnified views of the night skies. Innumerable deep-sky treasures can be viewed in detail and wonderful structure is visible on the Moon and planets. Viewed through a telescope, the stars themselves appear brighter but they are so far away that they remain pinpoints of light, regardless of the magnification used.

Spectacular images from the Hubble Space Telescope and a bewildering assortment of satellites and space probes show a magnificently pin-sharp, multicoloured Universe. It's easy to see why some people's expectations are incredibly high when they put their eye to the telescope eyepiece for the first time. While a great many celestial sights are truly amazing visually, in practical terms much of what the night skies has to offer its Earth-based viewers is relatively faint and requires a degree of informed appreciation. Cosmic colours are generally on the subtle side and the eye has to get used to its own limitations, as well as the limitations of the instrument used and those imposed by the local environment.

The enjoyment, however, is in the self-discovery of the night skies, learning what each celestial object is and where it's located in the Universe. The realization that your eyes are receiving rays of light from a distant object that may have set out before you were born, before Rome was built or even before humans evolved on Earth, is truly awe-inspiring.

Comparisons of views through different sized telescopes at the same magnification: the Trifid Nebula, as seen through a 60mm refractor (left), a 200mm Schmidt-Cassegrain (centre) and a 500mm reflector (right).

In keeping with the deep human desire to find some sort of order in the cosmos, patterns of stars in the night skies were assembled into constellations – creatures, objects and symbols outlined by prominent stars in a join-the-dots fashion. Constellations reflected the mythology and lifestyle of each culture that imagined them. Such celestial picture books had more than poetic purposes; agricultural communities used their rising, culmination and setting for timekeeping, while navigators and explorers found them to be useful signposts in the sky.

From the Sumerian and Babylonian civilizations of Bronze Age Mesopotamia arose the origins of the patterns of constellations we recognize today; they defined the ecliptic (the yearly path of the Sun), the 12 divine constellations of the Zodiac along the ecliptic (through which the Moon and planets appear to move), and numerous other constellations that referred to animals and agriculture.

These ancient constellations were later incorporated into works by Eudoxus of Cnidus (*c.*408–347BC), who devised a complete system of the Universe, envisaging Earth at the centre of a series of nested transparent crystal planets. Using naked-eye sighting devices, Hipparchus (*c.*190–120BC) made detailed observations of star positions, enabling him to create the first known star catalogue. It featured 48 classical constellations and around 850 stars whose position on the celestial sphere was pinpointed according to a system of celestial coordinates.

A magnitude scale devised by Hipparchus denoted each star's apparent brightness; the brightest 20 stars were classed as being of the first magnitude, followed by the next brightest which were second magnitude, and so on, down to the faintest stars, which he classed as sixth magnitude. A similar scale of star brightness is used today, although each division between magnitudes corresponds to a precise jump in brightness by a factor of 2.512 (gauged by photoelectric means).

Some time later, Claudius Ptolemy (*c.*90–168AD) compiled the *Almagest*, which used and expanded upon Hipparchus's work by producing a definitive atlas of the stars – 1022 of them, contained within the 48 classical constellations, themselves grouped into northern, Zodiacal and southern constellations.

Much of our knowledge of Greek philosophy – including the work of Hipparchus and Ptolemy – comes from ancient texts that were translated, copied and preserved by Arab scholars in Baghdad during the European Dark Ages. Abd al-Rahman al-Sufi (Azophi, 903–86), one of the greatest Arabic astronomers, produced *The Book of the Fixed Stars*, his own version of Ptolemy's star catalogue in which many of the stars were given Arabic names. Many of these names (albeit in modified form) remain in use to this day.

Sunrise over the Parthenon temple on the Acropolis hill in Athens. Built in the 5th century BC, the Parthenon is astronomically aligned with the rising of the Pleiades star cluster in Taurus.

PTOLEMY'S 48 CONSTELLATIONS

Andromeda (Andromeda, a princess)

Aquarius* (the Water-Carrier)

Aquila (the Eagle)

Ara (the Altar)

Argo Navis** (the Argo, a ship)

Aries* (the Ram)

Auriga (the Charioteer)

Boötes (the Herdsman)

Cancer* (the Crab)

Canis Major (the Great Dog)

Canis Minor (the Lesser Dog)

Capricornus* (the Goat)

Cassiopeia (Cassiopeia, a queen)

Centaurus (the Centaur)

Cepheus (Cepheus, a king)

Cetus (the Whale)

Corona Australis (the Southern Crown)

Corona Borealis (the Northern Crown)

Corvus (the Crow)

Crater (the Cup)

Cygnus (the Swan)

Delphinus (the Dolphin)

Draco (the Dragon)

Equuleus (the Little Horse)

Eridanus (the River Eridanus)

Gemini* (the Twins)

Hercules (Hercules, a hero)

Hydra (the Water Snake)

Leo* (the Lion)

Lepus (the Hare)

Libra* (the Scales)

Lupus (the Wolf)

Lyra (the Lyre)

Ophiuchus (the Serpent Holder)

Orion (Orion)

Pegasus (Pegasus, the winged horse)

Perseus (Perseus, a hero)

Pisces* (the Fishes)

Piscis Austrinus (the Southern Fish)

Sagitta (the Arrow)

Sagittarius* (the Archer, a centaur)

Scorpius* (the Scorpion)

Serpens (the Serpent)

Taurus* (the Bull)

Triangulum (the Triangle)

Ursa Major (the Great Bear)

Ursa Minor (the Little Bear)

Virgo* (the Virgin, a goddess)

*A Zodiacal constellation
**This was later split into three constellations – Carina (the Keel), Puppis (the Poop Deck) and Vela (the Sails).

Ancient Greek ideas eventually found their way back into the European arena during the High Middle Ages, as the works preserved by Arab scholars were translated into Latin. In the 16th century, Europe saw an explosion of scientific and astronomical enquiry when ancient explanations of the Universe were questioned and found wanting.

In his book *De Revolutionibus Orbium Coelestium* (*On the Revolutions of the Celestial Spheres*) Nicolaus Koppernik (Copernicus, 1473–1543) promoted the heliocentric theory – a model that places the Sun, not the Earth, at the centre of the Universe. Tycho Brahe (1546–1601), the last and greatest observer of the pre-telescopic era, made precise measurements of the stars and the movements of the planets using naked-eye quadrants and cross-staffs. Using Tycho's data, his student Johannes Kepler (1571–1630) placed Copernicus's heliocentric theory on a firm scientific footing.

Johann Bayer (1572–1625) used Tycho's star positions to produce much of the *Uranometria*, the first star atlas to cover the entire celestial sphere; the far southern stars, uncharted by classical scholars, were mapped according to the catalogue of the navigator Pieter Keyser (c.1540–96). *Uranometria's* 51 charts contain more than 2,000 stars, and 12 new constellations were allocated to the deep southern skies.

Importantly, *Uranometria* introduced the system of identifying the brighter stars in each constellation (down to the sixth magnitude) with letters of the Greek alphabet – Alpha being the brightest, Beta the second brightest, and so on. In some large constellations the Greek letters ran out, so Bayer used Roman letters, starting with a uppercase A followed by lowercase b, c, d, and so on. The system, devised before the telescope was invented, was neither precise nor perfect, but for all its idiosyncrasies it is retained to this day in much the same form as it originated.

THE GREEK ALPHABET (WITH SYMBOLS)

Alpha α / Beta β / Gamma γ / Delta δ / Epsilon ε / Zeta ζ / Eta η / Theta θ / Iota ι / Kappa κ / Lambda λ / Mu μ / Nu ν / Xi ξ / Omicron o / Pi π / Rho ρ / Sigma σ / Tau τ / Upsilon υ / Phi ϕ / Chi χ / Psi ψ / Omega ϖ

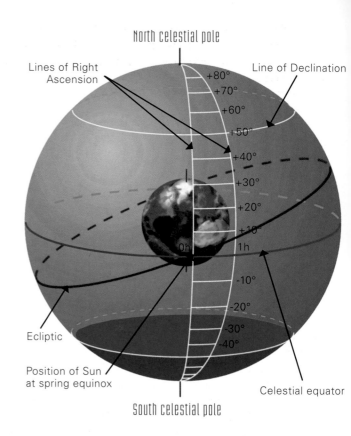

The celestial sphere, showing the north and south celestial poles, the celestial equator and the ecliptic. The shaded area around the south celestial pole is the region uncharted by the star-mappers of the ancient world.

Undeniable proof of a radically different layout of the Universe came with the invention of the telescope in the early 17th century. Galileo Galilei (1564–1642) discovered four satellites orbiting Jupiter, and Venus's phases showed that it was a globe in orbit around the Sun. It became increasingly obvious that Earth was just another planet with a satellite, orbiting the Sun between Venus and Mars.

Galileo discovered that the faint band of the Milky Way was made up of countless stars that were only visible through the telescope. It was reasoned that if the stars themselves were like the Sun – but so distant that they appeared as points of light – then perhaps the Sun wasn't so special. Instead of lying at the hub of the Universe, the Sun was found to be just one of a broader mass of stars making up the Milky Way.

Great advances were made in our understanding of the Universe through telescopic observation during the 17th, 18th and 19th centuries. As telescopes grew larger, familiar objects became better known, and new objects loomed into view from deeper, darker depths of the cosmos. Telescopic surveys of the skies charted the stars and catalogued deep-sky objects – star clusters and faint misty patches known as nebulae. From his private observatory, Johannes Hewelke (Hevelius, 1611–87) made accurate measurements of star positions and produced the most advanced star atlas of its time, the *Uranographia*. He devised a number of new constellations to lie among the traditional ancient Greek constellations. Then, using observations made from his own private observatory in England, John Bevis (1693–1771) compiled *Atlas Celeste*, an 18th century star atlas containing elaborately engraved star charts.

The 18th century finally saw the beginnings of the proper charting of the deep southern skies – a portion of the celestial sphere that had never been surveyed accurately before. Nicolas Lacaille (1713–62) set up an observatory at the Cape of Good Hope, from where he catalogued nearly 10,000 southern stars and 42 deep-sky objects. Lacaille's *Coelum Australe Stelliferum* introduced 14 new constellations in and around Bayer's deep southern constellations, all of which became accepted by the astronomical community.

Charles Messier (1730–1817) compiled a list of 110 deep-sky objects. Incorporating all sorts of objects that appear 'fuzzy' through a small telescope, Messier's list is an eclectic mix of star

Hevelius's chart of Taurus from the *Uranographia*.

clusters, nebulae and galaxies. It proved so useful that it is still referred to today by stargazers.

Using telescopes that he had made himself, William Herschel (1738–1822) surveyed the skies and recorded hundreds of double stars and nebulae. After discovering Uranus in one of his routine star-sweeps, Herschel established himself as the world's most prolific astronomer and went on to make many discoveries about the Solar System. William's sister, Caroline Herschel (1750–1848) was also a prolific observer of the skies, while his son, John Herschel (1792–1871) surveyed the skies of the southern hemisphere, recording hundreds of previously unseen double stars and nebulae.

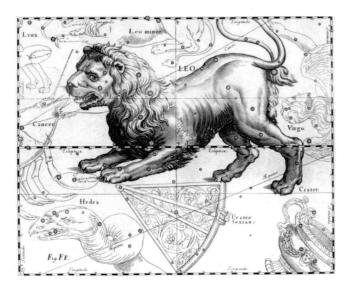

The constellation of Leo, from Hevelius's *Uranographia*.

As burgeoning European empires extended their reach around the globe, the increasing importance of accurate navigation by the stars led to the founding of national observatories, such as the Paris Observatory, France (1671), the Royal Greenwich Observatory in England (1675) and Germany's Berlin Observatory (1700). Telescopes were used to note the exact time that stars transited the meridian (appeared due south), enabling coordinates to be measured precisely.

From Greenwich the first Astronomer Royal, John Flamsteed (1646–1719) catalogued 2,935 stars and produced the most accurate celestial atlas of its day, the *Atlas Coelestis*; this was updated by Nicolas Fortin (1750–1841) in Paris, producing the smaller, more popular *Atlas Fortin-Flamsteed*, which had an artistic makeover and a number of newly discovered nebulae. In Berlin, Johann Bode (1747–1826) produced the *Vorstellung der Gestirne* (intended for amateur astronomers) and the *Uranographia*, which contains more than 17,000 stars and deep-sky objects and is widely regarded as the greatest pictorial star atlas of all time.

By the late 19th century, photography had advanced to such a state that it was possible to chart the skies on photographic plates. An international consortium established the *Carte du*

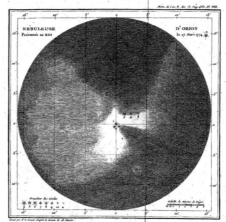

Charles Messier's drawing of the Orion Nebula (M42) compared with the author's observation of the same object.

Ciel, a project intended to chart the entire sky photographically using a number of identical astrographic telescopes set up around the world. It was an enormous project – more than 22,000 photographic plates were taken between 1881 and 1950 – and the task of physically measuring star positions down to the eleventh magnitude and with an accuracy of 0.5 arcseconds on each plate was incredibly laborious. The project was superseded in the 20th century using the very large wide-angle 48-inch Schmidt telescope at Mount Palomar in California; completed in 1958, the Palomar Sky Survey covers the sky from the north celestial pole to a declination of -30 degrees and shows stars down to the twenty-second magnitude on average (a million times fainter than those visible with the naked eye).

With improvements in optics and photography came wave after wave of deep-space discovery. Catalogues of newly discovered deep-sky objects were created to keep a tally, notable among which were John Dreyer's *New General Catalogue* (NGC, 1888) containing nearly 8,000 objects, and its appendices, the *Index Catalogue* (IC, 1895) with more than 5,000 objects. Many hundreds of NGC and IC objects are visible through an average-sized amateur telescope and they remain essential lists for today's astronomer.

The late 20th century saw the introduction of orbiting observatories high above Earth's turbulent atmosphere. Foremost among these, in terms of mapping the skies, was the European Space Agency's Hipparcos (High Precision Parallax Collecting Satellite). Between 1989 and 1993 Hipparcos precisely measured the positions of celestial objects, leading to the production of the *Tycho-2 Catalogue*, which contains 2.5 million stars. Data for stars nearby in our Galaxy has allowed astronomers to determine accurately the actual motion of stars through space and to gauge their distance from us using the parallax effect resulting from Earth's orbit around the Sun.

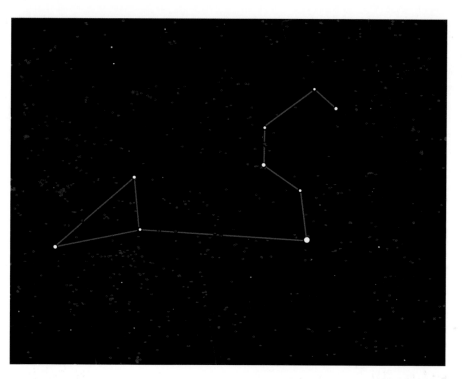

The constellation of Leo, showing all of its NGC objects – most of them are distant galaxies.

Even though the stars are at unimaginable distances from us, astronomers know a great deal about them and their life cycles. A star's mass is the overriding factor in determining how big it is and how brightly it shines throughout its life, how it develops during its lifetime and how long it lives.

All stars begin their lives in a gravitationally contracting cloud of interstellar dust and gas – mainly hydrogen and helium – within a galaxy. These clouds are so dense and deep as to be virtually opaque to visible light, and some of them can be seen silhouetted against parts of the Milky Way. Numerous stellar birthplaces may appear within each contracting zone of interstellar dust and gas. It may take around ten million years from the first stages of collapse to the appearance of an embryonic star, a region of unstoppable gravitational collapse called a protostar.

Dust and gas attracted by the protostar's gravity produces ever-rising heat and pressure at its core. Temperatures eventually become high enough to trigger thermonuclear fusion, where two hydrogen atoms combine at high speed to produce one helium atom, giving off a blast of energy, which ignites the star.

Much of the dust and gas surrounding the young star is blown away by the strong stellar wind, but any remaining material in orbit bathes in its new sun's light and energy. Astronomers have been able to detect and image such disks of gas and dust around newborn stars: known as proplyds (protoplanetary disks), they are solar systems in the making. Many beautiful proplyds have been imaged in the Orion Nebula, and among notable naked-eye stars with proplyds are Beta Pictoris, 51 Ophiuchi, Fomalhaut, Vega and Zeta Leporis.

All stars are not born equal. The amount of time a star spends fusing hydrogen into helium – living as a Main Sequence star – depends on its mass. The larger its mass, the hotter the star; its energy production is faster as its consumption of hydrogen fuel takes place at a faster rate, so it enjoys less time on the Main Sequence.

Perhaps the most well-known dark nebula is the Horsehead Nebula in Orion – a remarkably shaped projection of the edge of a larger molecular cloud, which is silhouetted against a glowing red gas cloud in the vicinity of the star Sigma Orionis. 1,500 light years away, the Horsehead Nebula is around ten light years in length. Inset, a closer Hubble Space Telescope view, where a young star, still swaddled in its nursery, can be seen shining at top left.

An 'average' star like the Sun, now around 4.7 billion years old, will spend another five billion years on the Main Sequence before entering its old age. Huge stars with 50 times the Sun's mass will spend just a million years on the Main Sequence before using up all their hydrogen fuel.

The smallest stars on the Main Sequence – brown and red dwarfs with a mass less than one-third of the Sun – have convective interiors with hydrogen and helium mixed throughout. This allows them to burn up a far greater proportion of their hydrogen fuel than more massive stars, giving them exceedingly long lives of many tens of billions of years – longer than the current age of the Universe. As a consequence, no red dwarf in the late stages of its life has ever been observed.

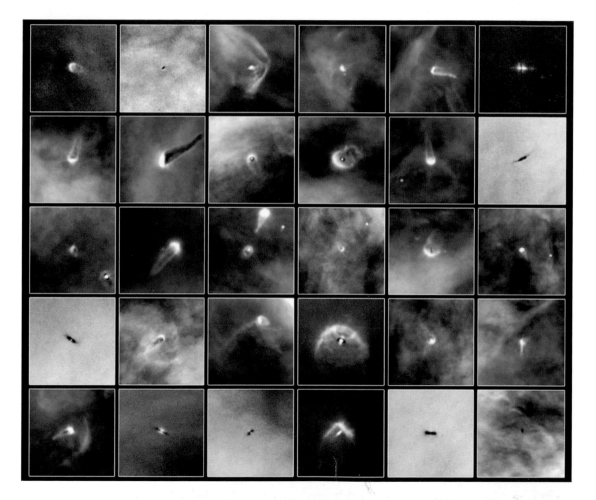

A fantastic assortment of proplyds, imaged in the Orion Nebula by the Hubble Space Telescope.

A star's colour gives us an idea of its surface temperature. Redder stars are cooler, while bluer stars are hotter. Our own yellow Sun has a surface temperature of around 6,000°C (10,800°F). Betelgeuse (Alpha Orionis) – an old red supergiant in the final stages of its life, 1,000 times the size of the Sun and 600 light years away – has a surface temperature of around 2,000°C (3,600°F), which is about one-third of the Sun's temperature.

It was once maintained that the composition of stars would forever remain a mystery. This was proved wrong when the spectroscope – a special piece of equipment used to split light into a spectrum for analysis – clearly revealed chemical elements in the stars. A multitude of thin dark lines in stellar spectra, known as absorption lines, are like the fingerprints of the elements.

Spectroscopy allows the directional speed of stars to be determined by their 'Doppler shift', a line-of-sight effect that displaces absorption lines towards the red or blue end of the spectrum; the degree to which light is shifted allows its relative speed to be gauged. Redshifted light is produced by a star moving away from the observer, while an approaching star will show a blueshift.

The spectral types are identified by seven main groups – O, B, A F, G, K and M – from hot O-type blue stars, through G-type stars like the Sun, to cool M-type red stars.

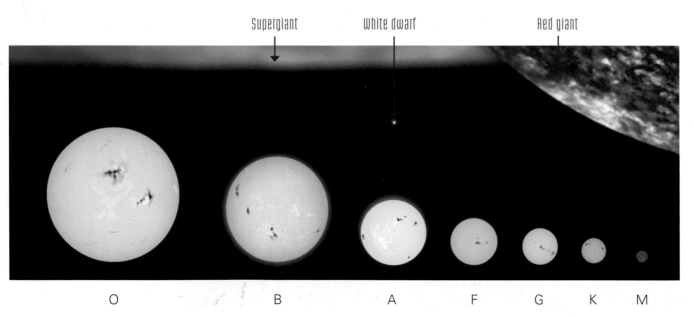

Supergiant · White dwarf · Red giant

O · B · A · F · G · K · M

Stars come in a vast range of sizes. Main Sequence stars, from hot O-types, through G-types like the Sun, to M-type red dwarves, compared to off-sequence stars like tiny white dwarf remnants and huge old red giants and supergiants.

It was once thought impossible that planets around other stars (exoplanets) could be detected as stars are so incredibly distant that any planets would appear so close to them as to be unresolvable as individual objects. Moreover, as planets shine only by reflected light, they were considered far too faint to detect, their light being drowned out by starlight. Thanks to ultra-sensitive measurements using the latest technology, the age-old dream of finding exoplanets has now finally been realized.

One technique for detecting exoplanets measures minute variations in a star's velocity (its Doppler shift) caused by the gravitational tug of orbiting planets; the closer and more massive the planet to its parent star, the larger the observed effect. This technique can only tell us the minimum size of the exoplanet, its orbital period and distance from the star.

Another detection technique measures variations in a star's brightness caused when, from our viewpoint, a planet passes in front of it; silhouetted against the star, the transiting planet causes a tiny but detectable drop in brightness. This technique reveals a great deal more, including the mass and radius of both the star and planet. Although only a small proportion of stars and their planetary systems are aligned in our line of sight, the transit technique allows astronomers to gauge an exoplanet's gravity and density, and to speculate reasonably about its composition and surface conditions.

It is now thought that a large proportion of stars have planetary systems, including around half of all Sun-like stars. In our Galaxy alone there are likely to be many tens of billions of planets. In 1995 the star 51 Pegasi became the first Sun-like star known to have an exoplanet – one with half the mass of Jupiter and orbiting at a blistering pace of once every 4.2 days. Since then, hundreds of exoplanets have been discovered in a wide variety of exoplanetary systems.

Exoplanets have been discovered using both Earth-based telescopes and orbiting observatories, notably NASA's Kepler Mission, which was launched in 2009 specifically to discover Earth-like exoplanets in our galactic neighbourhood. Once in orbit, NASA's James Webb Space Telescope will be able to follow up Kepler's amazing discoveries and measure the chemical composition of transiting exoplanets' atmospheres with great accuracy. This is an important step in finding out whether our own planet – orbiting in a 'warm' zone allowing liquid water to exist on its surface and just the right size to hold on to a life-friendly atmosphere – is rare among the stars, or whether conditions amenable to the development of life exist in abundance in the Universe.

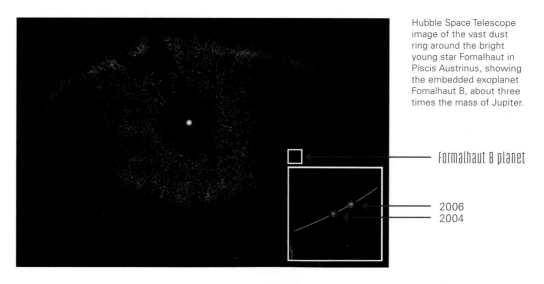

Hubble Space Telescope image of the vast dust ring around the bright young star Fomalhaut in Piscis Austrinus, showing the embedded exoplanet Fomalhaut B, about three times the mass of Jupiter.

Fomalhaut B planet

2006
2004

A star eventually uses up the hydrogen fuel at its core; energy production falls and the core's temperature and pressure decreases. In response, the core contracts slightly, producing a sharp rise in temperature which in turn ignites the hydrogen shell surrounding the core – a zone that was previously too cool to undergo fusion reactions. At this point the star leaves the Main Sequence; as it expands, the star's increased surface area makes it appear brighter, while its surface becomes cooler and redder. The star has become a red giant.

As the star's core contracts yet further and pressure increases, temperatures rise high enough to begin burning the remaining helium within the core, fusing it into carbon. As the core periodically contracts, the star's outer atmosphere is puffed into space as rings or shells of material; they are known as planetary nebulae because some resemble ghostly planets when viewed through a telescope. At their centre lies the highly compressed million-degree stellar remnant – an Earth-sized object called a white dwarf, so dense that a thimble full of its stuff would weigh a metric ton. Hydrogen gas within planetary nebulae glows as it is ionized by ultraviolet radiation emitted by the white dwarf. But planetary nebulae are short-lived – they are only visible for less than 100,000 years, as they expand and fade. Around 1,000 planetary nebulae populate our part of the Galaxy.

Although they shine for only a brief moment on the cosmic stage, planetary nebulae are among the most beautiful objects in the Universe. Although they have all been formed in the same way, these puffs of gas from dying stars vary enormously in appearance. Easily the largest and brightest of these, the Dumbbell Nebula in Vulpecula, is visible through binoculars; it has two brightly glowing lobes, giving it the appearance of a luminous apple core. Another lovely bright planetary nebula, the Ring Nebula (M57) in Lyra, is a beautiful telescopic sight – a luminous doughnut whose central white dwarf is just visible through larger instruments.

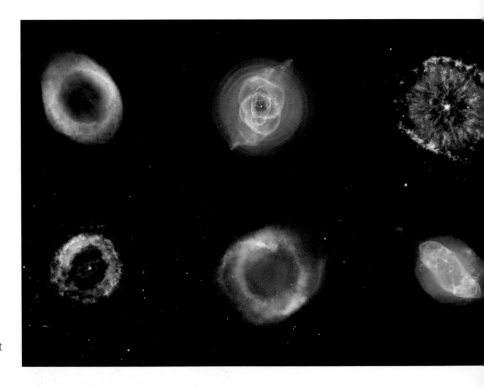

A selection of planetary nebulae imaged by the Hubble Space Telescope, all of which are visible though a telescope. Clockwise from top left: the Ring Nebula (M57) in Lyra, the Cat's Eye Nebula (NGC 6543) in Draco, the Glowing Eye Nebula (NGC 6751) in Aquila, the Saturn Nebula (NGC 7009) in Aquarius, the Helix Nebula (NGC 7293) in Aquarius and the Little Ghost Nebula (NGC 6369) in Ophiuchus.

After transforming into red supergiants, really massive stars refuse to fade gently into oblivion. With their nuclear fuel exhausted at the end of their lives, they become highly unstable and their cores eventually collapse. Within a fraction of a second, the core – an object the size of the Sun – crumples in on itself under its own gravity. A shockwave rebounds through the star's outer layers and a supernova ensues – a catastrophic explosion that produces so much light that it can briefly outshine all the stars in its home galaxy. Known as supernovae, these brief events on the cosmic timescale can produce as much energy as the Sun emits during its entire lifetime.

At the centre of the explosion the core's collapse may stop when it is squeezed into a neutron star – an object with about the same mass as the Sun but packed into an ultra-dense object just 10km (6 miles) across. When first formed, neutron stars spin at incredibly fast rates, and they emit radio waves along their rotation axis – these are known as pulsars because radio telescopes can detect regular pulses of radio emission from them.

The cores of stars more massive than three Suns are crumpled up so thoroughly by gravity during a supernova explosion that nothing can stop the process. Their gravity becomes so great that nothing can escape from it – not even light. Black holes have effectively opted out of the Universe – the known laws of physics stop at a black hole's edge, known as the event horizon. Most galaxies, including our own, are thought to have a supermassive black hole at their centre, ranging in size from hundreds of thousands to billions of solar masses.

In 1054AD, a brilliant new star suddenly appeared in the constellation of Taurus. It was bright enough to be visible in daylight for three weeks and remained a naked-eye object for almost two years as it gradually dimmed. Still visible through a small telescope as a faint glowing patch, and shown here in detail by the Hubble Space Telescope, the remnant of this supernova explosion is called the Crab Nebula (M1) in Taurus. Radio waves and visible light from the super-dense, rapidly spinning pulsar at its centre are flashed our way 30 times a second.

Our galactic near neighbour, the Large Magellanic Cloud, played host to a supernova explosion in February 1987. This Hubble Space Telescope image shows the supernova remnant, surrounded by beautiful glowing rings, local dust and gas heated up by the rapidly propagated supernova shockwave.

Venturing beyond the Solar System – our modest enclave nestled within a spiral arm of the Milky Way Galaxy – we enter the awe-inspiring realms of deep space. Thousands of deep-sky objects, including coloured and multiple stars, star clusters and glowing nebulae, can be viewed through binoculars and small telescopes. Yet more deep-sky treasures lie far beyond the Milky Way, thousands of which are revealed at the telescope eyepiece.

Stars aren't always single. In fact, around half the stars in our Galaxy have one or more close stellar siblings – double or multiple stars so close to each other that they orbit a common centre of gravity. Many stars that appear to be close together are line-of-sight doubles; their apparent proximity to each other results from the angle at which we view them. Such doubles are so far apart that they have no gravitational influence over each other.

Victorian astronomers took particular pleasure in identifying double stars, measuring their separations and noting their colours. Modern stargazers can still take great pleasure from observing them – especially doubles with prominent colours or striking colour contrasts.

Mizar, the second star from the end of Ursa Major's tail, is perhaps the most famous naked-eye double star. Keen-sighted people will be able to discern its fainter partner, Alcor, without optical aid. Known to the English as 'Jack and his Wagon' in years gone by, the pair are relatively close together in space, separated by just three light years. A small telescope reveals that Mizar has a much closer partner than Alcor – Mizar B; this pair was the first double star ever discovered through the telescope, way back in 1650. Alcor takes around a million years to make one circuit of the Mizar system, while the closer component orbits Mizar every 5,000 years.

Epsilon Lyrae, another popular bright double star, can be separated into two without optical aid by those with excellent eyesight. Both the stars in this binary has its own partner, making it a 'double-double' star, each component of which can be resolved through a small telescope at a high magnification. It's also worth mentioning Theta Orionis (the Trapezium). Located at the heart of the spectacular Orion Nebula, this lovely compact group of young, hot stars illuminates its gaseous neighbourhood. Small telescopes show the Trapezium's four brightest stars surrounded by a greenish nebulosity.

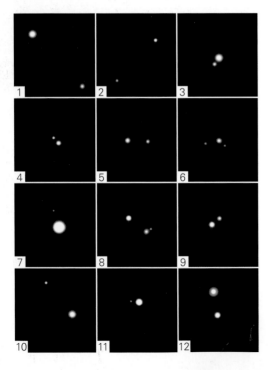

Dozens of lovely coloured double stars can be viewed through a small telescope, and some of the brightest and most beautiful examples are featured here (colours exaggerated for clarity). 1. Beta Cygni (Albireo); 2. Eta Persei; 3. Epsilon Boötis; 4. Xi Boötis; 5. Gamma Delphini; 6. Iota Cassiopeiae; 7. Beta Orionis (Rigel); 8. Upsilon Andromedae; 9. Alpha Herculis; 10. Alpha Canum Venaticorum (Cor Caroli); 11. Alpha Scorpii (Antares); 12. Beta Scorpii.

Bright stars sometimes appear to twinkle or even flash an array of colours – this is caused by unsteadiness in the Earth's atmosphere and has nothing to do with the output of light from the stars themselves. However, many stars really do vary in brightness over a longer period of time, either on a regular, semi-regular or irregular basis. Some stars show only small variations in their brightness, while others undergo major fluctuations ranging over many orders of magnitude.

Eclipsing binaries

Known by its ancient Arabic name of Algol (the 'demon'), the second brightest star in Perseus experiences a substantial dip in brightness every 2.87 days, remaining at minimum brightness for around 10 hours. Algol is a type of variable star known as an eclipsing binary. Its dip in brightness – easily noticeable with the unaided eye – is caused when it is temporarily eclipsed by a much larger, dimmer star orbiting around it.

There are many more examples of eclipsing binary stars. One of the most fascinating of these is the star Epsilon Aurigae, the apex star in the small triangle asterism of the Kids near the bright star Capella. Every 27.1 years Epsilon, a yellow-white supergiant star, is eclipsed for a period of two years by an unseen star surrounded by a thick ring of dust.

Cepheids

A certain class of supergiant stars known as Cepheid variables rhythmically expand and contract, varying in brightness with periods varying between 2 to 50 days. Because their brightness is related to their cycle of variability, Cepheids are immensely valuable in determining the size of the Milky Way and the nearness of other galaxies – the brighter the star, the longer its period. By comparing a Cepheid's apparent and actual brightness, its distance can be accurately gauged.

Algol is observed to fluctuate in brightness on a regular basis when a large, dim orbiting companion periodically eclipses the star.

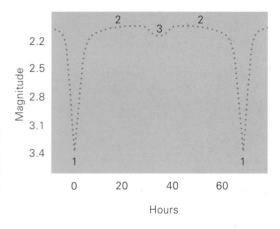

Phases of the eclipse: minimum (1), maximum (2) and secondary dip (3) in brightness.

Long-period variables

Slowly pulsating long-period variable stars expand and contract to produce changes in luminosity, just like Cepheids, but their cycles of variation follow only a general pattern. The best known long-period variable, Mira (the 'wonderful') in the constellation of Cetus, has an average period of around 330 days during which it ranges from being easily visible with the unaided eye to being difficult to detect through binoculars.

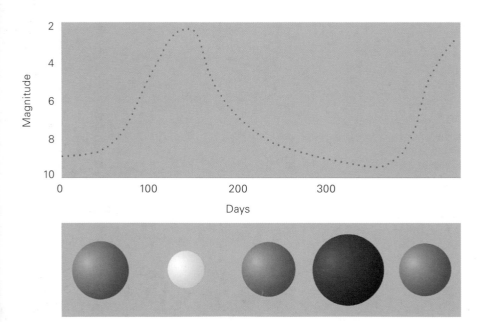

Mira, a long-period variable star, fluctuates in size and brightness over a period of around 330 days, as shown in these two diagrams.

Irregular variables

Stars with erratic changes in their brightness and/or period are known as irregular variables. There are two types – pulsating irregulars and eruptive irregulars. Pulsating irregulars are old supergiant stars approaching the end of their life, expanding and collapsing unpredictably as their nuclear fuel runs low. Eruptive irregulars undergo sudden increases in brightness as they eject material into space, their surfaces temporarily brightening to many times their previous magnitude.

Stellar cataclysms

Cataclysmic variable stars usually consist of a white dwarf primary and a red dwarf secondary. When a star in a closely orbiting binary star system leaves the Main Sequence towards the end of its life to become a white dwarf, its less massive red dwarf partner, still on the Main Sequence, finds itself orbiting a stellar cannibal. The pair is so close that the white dwarf's gravity distorts the shape of its neighbour, and its gaseous mantle of hydrogen is dragged away to form a hot accretion disk around the white dwarf. Occasional infalls of hydrogen gas onto the surface of the greedy white dwarf produces a sudden outburst – a nova (Latin for 'new'), so called because the star increases in brightness so much that it gives the appearance that a new star has suddenly appeared from nowhere.

Shaped like a vast flat disk with a bulging centre, the Milky Way galaxy is 100,000 light years across and contains at least 200 billion stars. Contained within the Galaxy's centre is a tightly packed population of ancient, low-mass red stars. In the surrounding flattened galactic disk, active star-forming regions illuminate gracefully sweeping spiral arms; contained within them are brilliant young white and blue stars and their associated nebulae. At first glance, the nucleus and brightly glowing arms of a spiral galaxy appear to constitute its entire fabric, but in fact, the dark lanes that separate a galaxy's bright spiral arms are opaque clouds of dust and gas, the material from which populations of future stars will be born.

Our own Solar System lies around two-thirds of the way from the galactic centre to its edge. From our position in the galactic plane, the stars of the Milky Way recede into the distance, creating a glowing band that circles the sky.

Most of the stars visible with the unaided eye – around 6,000 of them under the very best conditions – lie in our small cosmic patch. Only one-third of the 100 nearest stars shine brightly enough to be seen with the unaided eye. Of these, only three are nearer than ten light years distant. Light from the nearest naked-eye star, the Sun-like Alpha Centauri, takes just 4.37 years to reach us, while the furthest naked-eye star, the supergiant Iota-2 Scorpii, is 3,700 light years distant.

Every type of Main Sequence star can be found in our cosmic vicinity, from faint M-type red dwarfs like Proxima Centauri to O-type blue stars like Zeta Puppis. There's a good proportion of brown dwarfs and a sprinkling of white dwarfs – stars near the end of their lives.

Barnard's Star, a dim red dwarf in Ophiuchus, is just visible through binoculars. Six light years distant, it is the second closest star to the Sun after the Alpha Centauri system. In 1916, astronomer Edward Barnard discovered that the star has the largest known proper motion of all known stars – it moves against the celestial sphere at the staggering rate of 10.3 arcseconds per year, or about half the apparent diameter of the Moon each century. Barnard's Star is approaching us at a relative velocity of 500,000km per hour (310,000mph), and in around 10,000 years' time it will make its closest approach to us, at a distance of 3.8 light years.

Sirius is the brightest star in the entire sky – twice as bright as any other star. Located just 8.6 light years away, this brilliant A-type blue-white star is the fifth closest to the Sun. Of course, Sirius appears so bright because of its nearness to the Sun – there are many more intrinsically brighter stars.

Seen from a distant galaxy, high above the plane of the Milky Way, our own Galaxy would look a lot like this Hubble image of galaxy UGC 12158 in Pegasus. The approximate position of our Solar System is marked with a red dot.

Stars are rarely born in isolation. Our own Sun, now a singleton, was likely to have had a number of sister stars within the same cloud of dust and gas from which it was born around five billion years ago. Clusters of newborn stars have been detected buried deep within galactic clouds of dust and gas, hidden from direct view but detectable by the heat that they emit in the form of infrared radiation.

More than 1,000 open star clusters are known within our galactic vicinity. Since our view of the Milky Way is largely obscured by dust and gas clouds in the galactic plane, this likely represents a tiny percentage of the true figure.

Several million years after a galactic star cluster is born, its dusty, gaseous cocoon is dispersed by the strong stellar winds blowing from its hot new offspring. Over a period of a few hundreds of millions of years, the weakly gravitationally bound stellar siblings become spread out along the galactic plane by wider gravitational forces.

Open star clusters offer astronomers valuable insights into stellar evolution. Their constituent stars are all around the same distance, so their brightness can be compared like for like without difficulty. Having been formed from the same cloud of dust and gas, stars within open clusters are all around the same age as each other and share similar chemical compositions. Open clusters often contain stars with a wide range of masses, from dwarfs smaller than the Sun to giants up to a hundred solar masses, so their varied stellar contents can give them a jewel-box like appearance, often sparkling with multicoloured gems.

The Milky Way contains an abundance of gas and dust clouds. Some are visible only because they appear in silhouette against a brighter starry background. Others shine by reflecting the light of nearby stars or by emitting light of their own.

Viewed under dark conditions, far from the light-polluted city, an amazing amount of detail can be seen along the length of the glowing band of the Milky Way. Much of the visible structure of the Milky Way is produced by vast clouds of interstellar dust and gas along the plane of the Galaxy, which hide the light of distant glowing starfields.

The beautiful 'Jewel Box' open cluster lies in the Southern Cross. Its brightest stars shine a variety of colours.

A great dark rift running through the Milky Way in the constellation of Cygnus – sometimes called the 'northern coalsack' – is produced by a dense galactic cloud several tens of thousands of light years long, separating our own spiral arm from the Cygnus arm of the Milky Way.

Reflection nebulae

Light from bright newborn stars illuminates the surrounding dust and gas of their cosmic wombs, producing reflection nebulae. Their brightness depends on the size and density of the grains of dust reflecting the starlight, and the colour, brightness and proximity of the stars that illuminate them. Reflection nebulae often have a pronounced blue colour, caused by the reflective properties of carbon dust gains (blue wavelengths being more easily scattered than red).

A dark spire of gas and dust extends for around ten light years across part of the Eagle Nebula, an active stellar birthplace. Some of the bumps on its periphery may be on their way towards condensing into stars. Ultraviolet light streaming from nearby hot massive young stars (beyond the top of the image) is pushing against the spire, eroding it.

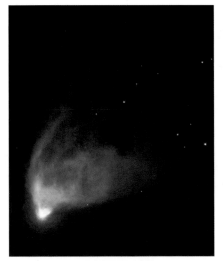

Hubble's Variable Nebula in the constellation of Monoceros is a broad fan-shaped reflection nebula whose brightness changes with the variations of its illuminating star R Monocerotis, which lies at its apex. Streamers of dust near the star cast giant shadows onto the nebula's walls, producing an ever-changing spectacle.

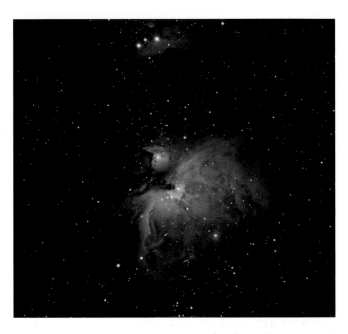

Of all the sky's emission nebulae, the Great Nebula in Orion is the brightest and most spectacular. Easily visible as a faint glowing patch in the sword handle asterism of Orion, a distinct green hue can be discerned through the eyepiece of an amateur-sized telescope. Photographs show a predominantly red nebula; delicate streamers of gas sweep gracefully away from the nebula's central region, which is punctuated by the four bright newborn stars of the Trapezium cluster. Nearby, a dark nebula intrudes against the brighter background glow, forming a feature known as the 'shark's mouth'. Observing it through his 1.2-metre reflecting telescope in 1789, William Herschel described it as 'an unformed fiery mist, the chaotic material of future suns.' Located some 1,600 light years away, the Orion Nebula is part of a far larger but much fainter diffuse nebula spanning virtually the whole of Orion.

Our Galaxy is surrounded by a vast halo of around 1,500 globular clusters – hefty star swarms consisting of hundreds of thousands of old red stars held together in a spherical mass by their mutual gravity. Globular clusters are extremely ancient entities, their stars having been among the first formed in the Galaxy. An average globular contains hundreds of thousands of old red stars and measures more than 100 light years across; their stars become more tightly packed towards their centres, the more crowded globulars having average star distances of only a few light weeks or months near their cores.

Astronomers have attempted to understand how these vast spherical stellar congregations came into being, and how their globular shapes have been maintained for billions of years. It is possible that giant black holes lurk at the centres of some globular clusters, providing the massive hub around which the cluster revolves. However, nothing so exotic may exist within most globulars; instead, their densely packed stars undergo an uneasy gravitational waltz.

Vast though our own Milky Way is, it is just one of billions of individual star systems scattered throughout the Universe. Some galaxies are far smaller, some appreciably larger than the Milky Way. Some take on graceful forms such as broad, curving spirals and smooth cigar-shaped ellipses, while others appear to have irregular shapes.

The Milky Way is part of a cluster of more than 50 galaxies known as the Local Group, a collection held together by mutual gravitational attraction. Among those nearest to us are two much smaller satellite galaxies known as the Small and Large Magellanic Clouds, visible as hazy patches with the unaided eye from the southern hemisphere. Two large members of the Local Group – the Andromeda Galaxy (M31) and the Pinwheel Galaxy (M33) in Triangulum – lie at distances of 2.5 and 2.8 billion light years respectively and both can be glimpsed with the naked eye under good conditions.

The Local Group is but one of a hundred similar galaxy clusters belonging to an even bigger gravitationally bonded entity – the Virgo Supercluster, spanning more than

M13, the largest globular cluster visible in northern skies, can be just located with the unaided eye in the constellation of Hercules. It is delightful to view telescopically, and many of its brighter stars can be resolved to its dense core, shown here up close in this image from the Hubble Space Telescope. It lies around 22,000 light years away and contains around 300,000 stars.

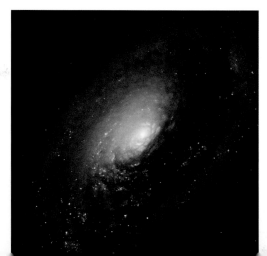

The Black Eye Galaxy (M64) in Coma Berenices is a prominent member of the Coma-Virgo cluster of galaxies.

a hundred million light years. To stretch our imagination yet further, there are millions of superclusters in the observable Universe.

As we look into the depths of intergalactic space, we look at increasingly earlier epochs in the history of the Universe. The nearest galaxy is 180,000 light years away, while the light from the most distant known galaxies has taken more than 13 billion years to cross the Universe to meet us.

By looking at galaxies over such a vast period of time, astronomers have gained an understanding of how such galaxies were created and how they evolve. Galaxies were formed within large gas clouds very early in the history of the Universe. Exactly how the galaxies were seeded is not completely understood, but studies of the early Universe suggest that the gas within it – hydrogen and helium – somehow developed ripples. In turn, knots of greater density collapsed under their own gravity to form the first star clusters and galaxies.

Incredibly, only one-tenth of the stuff making up the Universe can be directly detected – the remaining dark matter makes up a staggering 90 per cent of the mass.

Dark matter is known to exist because it produces a measurable gravitational effect on galaxies and causes a phenomenon called gravitational lensing, where light from more distant galaxies is bent, making them appear deformed or even producing multiple images of the same background objects. Some dark matter may be in the form of black holes, old neutron stars, brown dwarfs and dust, but the majority of it remains unknown.

Observations by Edwin Hubble (1889–1953) showed that the light from distant galaxies appears stretched out – or redshifted – towards the red end of the spectrum. Generally speaking, it was discovered that the further a galaxy is, the greater its redshift, implying that it is moving away at a greater speed. It's an inescapable conclusion that the Universe is expanding, as if flying from some unimaginable explosion in the distant past. Termed the Big Bang, this explosion is thought to have taken place around 13.7 billion years ago and was the beginning point of the Universe that we know. Somewhat counter-intuitively, the rate at which the Universe is expanding has been found to be increasing – this acceleration is being driven by a force known as dark energy whose nature is a mystery to science.

Galaxies often have a bumpy ride. Here, galaxies UGC 1810 and UGC 1813 are in the process of collision, both warping out of shape by their mutual gravitational tidal pulls. Image by the Hubble Space Telescope.

In this deep-field image taken by the Hubble Space Telescope, distant galaxies are bent into glowing arcs of light by gravitational lensing.

Since ancient times, the stars were imagined to be arrayed upon the inside of a vast celestial sphere at whose centre lay the Earth; moving between us and the sphere, the Sun, Moon and planets appeared projected against the starry background. Even though we know that the Universe is far different, we continue to use this analogy for positional astronomy because it is easy to work with.

Extensions of Earth's polar axis and equator define the position of the celestial poles and equator, and like

a terrestrial globe the celestial sphere has a simple system of coordinates.

Parallel lines of declination measured up to +90 degrees (north) and -90 degrees (south) from the celestial equator correspond to terrestrial latitude, while the celestial equivalent to longitude is represented by great circles of right ascension (RA), measured from 0 to 24 hours around the celestial equator. Each degree of declination and hour of RA is split into 60 arcminutes and each arcminute is further divided into 60 arcseconds.

Our annual orbit around the Sun produces another important line on the celestial sphere – the ecliptic – a line traced out by the Sun's apparent path among the stars, and the 12 traditional Zodiacal constellations all straddle the ecliptic. Because Earth's axis of rotation is tilted by 23.5 degrees – the ecliptic makes an angle of 23.5 degrees with the celestial equator. Earth and all the major planets orbit the Sun in almost the same plane so they follow paths close to the ecliptic. The phenomena of eclipses, produced when the Earth, Moon and Sun are in alignment, gives us the word ecliptic.

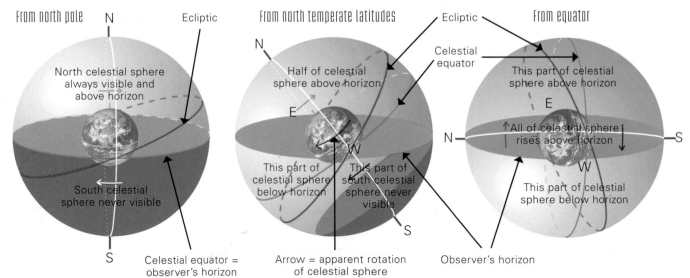

From north pole

N

Ecliptic

North celestial sphere always visible and above horizon

South celestial sphere never visible

S

Celestial equator = observer's horizon

From north temperate latitudes

Ecliptic

Celestial equator

N

Half of celestial sphere above horizon

E

W

This part of celestial sphere below horizon

This part of south celestial sphere never visible

S

Arrow = apparent rotation of celestial sphere

Observer's horizon

From equator

This part of celestial sphere above horizon

E

All of celestial sphere rises above horizon

N — S

W

This part of celestial sphere below horizon

Our location determines how much of the celestial sphere is seen annually. From the poles only half of the celestial equator is ever visible, but all of it can be seen from the equator over time. From temperate locations most of the celestial sphere comes into view except for a region around the opposite celestial pole.

Time and season

Rotating on its axis every 24 hours, the Earth orbits the Sun once every 365.25 days – the extra quarter day is added every four years to give us an extra day in February, in the form of a leap year.

While Earth's axis is tilted by 23.5 degrees to the plane of the ecliptic, the tilt is constant with respect to the stars; this gives us the seasons, whose effects are more noticeable further away from the equator. In December the Earth's north pole is angled away from the Sun, and during the northern winter the Arctic regions are in permanent darkness; meanwhile the southern hemisphere enjoys the height of summer, when the Antarctic is bathed in 24-hour sunlight. Precisely the opposite happens six months later when, during June, the Earth has moved to the other side of its orbit around the Sun. Now the north pole is angled towards the Sun at the height of northern summer, while it is midwinter in the southern hemisphere. These two extremes are called the winter and summer solstices. Between the solstices are the spring and autumn equinoxes, when neither pole is angled towards the Sun, and all parts of the Earth experience 12 hours of sunlight and 12 hours of darkness.

The stars' annual parade

From any particular part of the Earth there is a region surrounding the celestial pole that is always visible, and the extent of this circumpolar region varies with latitude. From either of the Earth's poles, exactly half the sky – everything above the horizon – appears circumpolar. Polaris (Alpha Ursae Minoris, known as the North Star because it is located close to the north celestial pole) is directly overhead at the north pole; as the Earth revolves the stars track parallel to the horizon, neither rising nor setting. From the latitude of London at 52°N, the north celestial pole is 52 degrees high; everything north of +38 degrees on the celestial sphere is circumpolar, while stars south of this declination appear to rise and set. Nothing south of -38 degrees on the celestial sphere ever rises from London. The same holds true for locations in the southern hemisphere.

From the latitude of Canberra (35°S), all stars south of -55 degrees are circumpolar, while everything north of +35 degrees lies permanently below the horizon.

Viewing non-circumpolar constellations depends largely on the season. The Sun's glare renders the constellations immediately surrounding it practically impossible to view. As the Sun proceeds along the ecliptic, formerly invisible constellations begin to emerge into the morning skies, reaching their highest above the horizon at midnight around six months later. Constellations then sink gradually into the evening skies, moving ever westward until they are once again lost in the evening afterglow and the glare of the Sun.

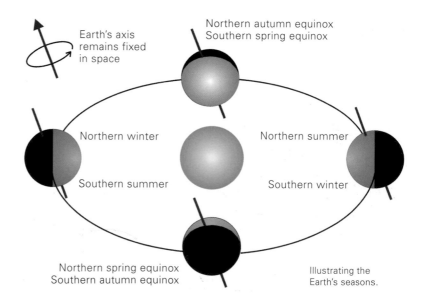

Earth's axis remains fixed in space

Northern autumn equinox
Southern spring equinox

Northern winter
Southern summer

Northern summer
Southern winter

Northern spring equinox
Southern autumn equinox

Illustrating the Earth's seasons.

HOW TO USE THIS BOOK ....

The star charts in this book show all the naked-eye stars (down to magnitude +5) visible under dark skies from both the northern and southern hemispheres. If a bright star isn't on the chart, it's more than likely to be one of the five bright planets – Mercury, Venus, Mars, Jupiter or Saturn, each of which can grow bright enough to be visible from urban locations.

Complete coverage of the celestial sphere is given by the ten wide-angle star charts featured in The Starry Skies section. All 88 constellations are depicted, labelled in abbreviated form. Each chart overlaps slightly with its neighbours, enabling them to be referred to more easily. Identification of the constellation patterns is made easier with the inclusion of joining lines. On each chart we highlight several of the more prominent constellations, accompanied by a tour of their bright stars, multiple stars, variable stars and deep-sky objects of interest; some of these are naked eye objects, while others require binoculars or a telescope. Reference is sometimes made to smaller, easily recognized patterns of stars such as the 'Keystone' of Hercules and the 'Plough' in Ursa Major; known as asterisms, these patterns are a great help in locating individual constellations and objects.

Instead of being an all-encompassing guide to the sky, our selective survey of the heavens serves as an ample taster for what the night skies have to offer the keen observer.

If you like what you see, there are many avenues along which you can pursue your interest in astronomy to a greater depth.

For practical purposes, each hemisphere begins with a chart showing the circumpolar constellations, followed by four seasonal views representing the sky above the southern horizon at midnight local time on 1 January (northern winter, southern summer), 1 April (northern spring, southern autumn/fall), 1 July (northern summer, southern spring) and 1 October (northern autumn/fall, southern spring). Representative horizon lines have been selected; for the northern hemisphere these are London (52°N) and New York (39°N), with Wellington (41°S) and Canberra (35°S) for the southern hemisphere. The band of the Milky Way features on the charts, and so too does the line of the ecliptic, near to which the Moon and planets can always be found.

The illustrations alongside each entry come from a variety of sources, but all of them are the work of amateur astronomers keen to capture the beauty of the night skies.

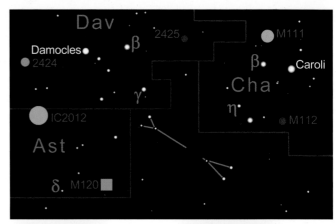

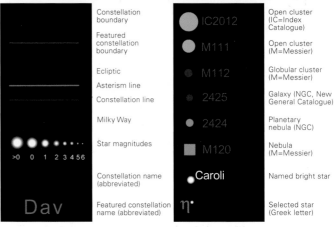

Constellation boundary		IC2012	Open cluster (IC=Index Catalogue)
Featured constellation boundary		M111	Open cluster (M=Messier)
Ecliptic		M112	Globular cluster (M=Messier)
Asterism line		2425	Galaxy (NGC, New General Catalogue)
Constellation line		2424	Planetary nebula (NGC)
Milky Way			
Star magnitudes >0 0 1 2 3 4 5 6		M120	Nebula (M=Messier)
Constellation name (abbreviated)		Caroli	Named bright star
Featured constellation name (abbreviated) Dav		η	Selected star (Greek letter)

Labelled hypothetical constellations, stars and deep-sky objects with explanatory key.

THE 88 OFFICIALLY RECOGNIZED CONSTELLATIONS

CONSTELLATION	ABBREVIATION	GENITIVE	MEANING	SIZE RANKING	BRIGHTEST STAR (MAGNITUDE)
Andromeda	**And**	**Andromedae**	**Andromeda (mythological character)**	**19**	**Alpheratz (2.1)**
Antlia	Ant	Antliae	Air pump	62	Alpha Antliae (4.3)
Apus	Aps	Apodis	Bird of paradise	67	Alpha Apodis (3.8)
Aquarius	**Aqr**	**Aquarii**	**Water bearer**	**10**	**Beta Aquarii (Sadalsuud) (2.9)**
Aquila	**Aql**	**Aquilae**	**Eagle**	**22**	**Altair (0.8)**
Ara	Ara	Arae	Altar	63	Beta Arae (2.8)
Aries	**Ari**	**Arietis**	**Ram**	**39**	**Hamal (2.0)**
Auriga	**Aur**	**Aurigae**	**Charioteer**	**21**	**Capella (0.1)**
Boötes	**Boo**	**Boötis**	**Herdsman**	**13**	**Arcturus (-0.1)**
Caelum	Cae	Caeli	Chisel	81	Alpha Caeli (4.4)
Camelopardalis	Cam	Camelopardalis	Giraffe	18	Beta Camelopardalis (4.0)
Cancer	**Cnc**	**Cancri**	**Crab**	**31**	**Beta Cancri (3.5)**
Canes Venatici	CVn	Canum Venaticorum	Hunting dogs	38	Cor Caroli (2.8–3.0 variable)
Canis Major	**CMa**	**Canis Majoris**	**Greater dog**	**43**	**Sirius (-1.4)**
Canis Minor	CMi	Canis Minoris	Lesser dog	71	Procyon (0.4)
Capricornus	**Cap**	**Capricorni**	**Sea goat**	**40**	**Algedi (2.7–2.8 variable)**
Carina	**Car**	**Carinae**	**Keel**	**34**	**Canopus (-0.6)**
Cassiopeia	**Cas**	**Cassiopeiae**	**Cassiopeia (mythological character)**	**25**	**Shedir (2.2)**
Centaurus	**Cen**	**Centauri**	**Centaur (mythological creature)**	**9**	**Rigil Kentaurus (-0.3)**
Cepheus	**Cep**	**Cephei**	**Cepheus (mythological character)**	**27**	**Alderamin (2.5)**
Cetus	**Cet**	**Ceti**	**Sea monster (later interpreted as a whale)**	**4**	**Diphda (Beta Ceti) (2.0)**
Chamaeleon	Cha	Chamaeleontis	Chameleon	79	Alpha Chamaeleontis (4.1)
Circinus	Cir	Circini	Compass (drawing tool)	85	Alpha Circini (3.2)
Columba	**Col**	**Columbae**	**Dove**	**54**	**Phact (2.7)**
Coma Berenices	**Com**	**Comae Berenices**	**Berenice's hair**	**42**	**Beta Comae Berenices (4.2)**

CONSTELLATION	ABBREVIATION	GENITIVE	MEANING	SIZE RANKING	BRIGHTEST STAR (MAGNITUDE)
Corona Australis	**CrA**	**Coronae Australis**	**Southern crown**	**80**	**Alpha Coronae Australis (4.1)**
Corona Borealis	**CrB**	**Coronae Borealis**	**Northern crown**	**73**	**Alphekka (2.2)**
Corvus	**Crv**	**Corvi**	**Crow**	**70**	**Gamma Coronae Borealis (Gienah) (2.6)**
Crater	Crt	Crateris	Cup	53	Delta Crateris (Labrum) (3.6)
Crux	**Cru**	**Crucis**	**Southern cross**	**88**	**Acrux (0.8)**
Cygnus	**Cyg**	**Cygni**	**Swan**	**16**	**Deneb (1.2)**
Delphinus	Del	Delphini	Dolphin	69	Beta Delphini (Rotanev) (3.6)
Dorado	Dor	Doradus	Goldfish	72	Alpha Doradus (3.3)
Draco	**Dra**	**Draconis**	**Dragon**	**8**	**Gamma Draconis (Etamin) (2.2)**
Equuleus	Equ	Equulei	Pony	87	Kitalpha (3.9)
Eridanus	**Eri**	**Eridani**	**River Eridanus (mythological)**	**6**	**Achernar (0.5)**
Fornax	For	Fornacis	Furnace	41	Alpha Fornacis (3.9)
Gemini	**Gem**	**Geminorum**	**Twins (Castor and Pollux, mythological characters)**	**30**	**Beta Geminorum (Pollux) (1.2)**
Grus	Gru	Gruis	Crane	45	Alnair (1.7)
Hercules	**Her**	**Herculis**	**Hercules (mythological character)**	**5**	**Beta Herculis (Kornephoros) (2.8)**
Horologium	Hor	Horologii	Pendulum clock	58	Alpha Horologii (3.9)
Hydra	**Hya**	**Hydrae**	**Hydra (mythological creature)**	**1**	**Alphard (2.0)**
Hydrus	Hyi	Hydri	Lesser water snake	61	Beta Hydri (2.8)
Indus	Ind	Indi	Indian (native American)	49	Alpha Indi (3.1)
Lacerta	Lac	Lacertae	Lizard	68	Alpha Lacertae (3.8)
Leo	**Leo**	**Leonis**	**Lion**	**12**	**Regulus (1.4)**
Leo Minor	LMi	Leonis Minoris	Lesser lion	64	46 Leonis Minoris (3.8)
Lepus	**Lep**	**Leporis**	**Hare**	**51**	**Arneb (2.6)**
Libra	**Lib**	**Librae**	**Balance**	**29**	**Beta Librae (Zubeneshamali) (2.6)**
Lupus	**Lup**	**Lupi**	**Wolf**	**46**	**Alpha Lupi (2.3)**
Lynx	Lyn	Lyncis	Lynx	28	Alpha Lyncis (3.1)
Lyra	**Lyr**	**Lyrae**	**Lyre (musical instrument)**	**52**	**Vega (0.0)**
Mensa	**Men**	**Mensae**	**Table Mountain (South Africa)**	**75**	**Alpha Mensae (5.1)**
Microscopium	Mic	Microscopii	Microscope	66	Gamma Microscopii (4.7)
Monoceros	**Mon**	**Monocerotis**	**Unicorn (mythological creature)**	**35**	**Beta Monocerotis (3.7)**
Musca	**Mus**	**Muscae**	**Fly**	**77**	**Alpha Muscae (2.7)**

CONSTELLATION	ABBREVIATION	GENITIVE	MEANING	SIZE RANKING	BRIGHTEST STAR (MAGNITUDE)
Norma	Nor	Normae	Carpenter's level	74	Gamma 2 Normae (4.0)
Octans	Oct	Octantis	Octant	50	Nu Octantis (3.7)
Ophiuchus	**Oph**	**Ophiuchi**	**Serpent-bearer**	**11**	**Rasalhague (2.1)**
Orion	**Ori**	**Orionis**	**Orion (mythological character)**	**26**	**Beta Orionis (Rigel) (0.2)**
Pavo	Pav	Pavonis	Peacock	44	Alpha Pavonis (1.9)
Pegasus	**Peg**	**Pegasi**	**Pegasus (mythological winged horse)**	**7**	**Epsilon Pegasi (Enif) (2.4)**
Perseus	**Per**	**Persei**	**Perseus (mythological character)**	**24**	**Mirphak (1.8)**
Phoenix	Phe	Phoenicis	Phoenix (mythological creature)	37	Ankaa (2.4)
Pictor	Pic	Pictoris	Easel	59	Alpha Pictoris (3.2)
Pisces	**Psc**	**Piscium**	**Fishes**	**14**	**Eta Piscium (Alpherg) (3.6)**
Piscis Austrinus	**PsA**	**Piscis Austrini**	**Southern fish**	**60**	**Fomalhaut (1.2)**
Puppis	**Pup**	**Puppis**	**Poop deck**	**20**	**Zeta Puppis (Naos) (2.2)**
Pyxis	Pyx	Pyxidis	Mariner's compass	65	Alpha Pyxidis (3.7)
Reticulum	Ret	Reticuli	Net (eyepiece graticule)	82	Alpha Reticuli (3.3)
Sagitta	Sge	Sagittae	Arrow	86	Gamma Sagittae (3.5)
Sagittarius	**Sgr**	**Sagittarii**	**Archer**	**15**	**Epsilon Sagittarii (Kaus Australis) (1.8)**
Scorpius	**Sco**	**Scorpii**	**Scorpion**	**33**	**Antares (0.9–1.2 variable)**
Sculptor	**Scl**	**Sculptoris**	**Sculptor**	**36**	**Alpha Sculptoris (4.3)**
Scutum	**Sct**	**Scuti**	**Shield (of Sobieski)**	**84**	**Alpha Scuti (3.8)**
Serpens	**Ser**	**Serpentis**	**Snake**	**23**	**Unukalhai (2.6)**
Sextans	Sex	Sextantis	Sextant	47	Alpha Sextantis (4.5)
Taurus	**Tau**	**Tauri**	**Bull**	**17**	**Aldebaran (0.8–1.0 variable)**
Telescopium	Tel	Telescopii	Telescope	57	Alpha Telescopii (3.5)
Triangulum	**Tri**	**Trianguli**	**Triangle**	**78**	**Beta Trianguli (3.0)**
Triangulum Australe	TrA	Trianguli Australis	Southern triangle	83	Atria (1.9)
Tucana	**Tuc**	**Tucanae**	**Toucan**	**48**	**Alpha Tucanae (2.9)**
Ursa Major	**UMa**	**Ursae Majoris**	**Great bear**	**3**	**Epsilon Ursae Majoris (Alioth) (1.8)**
Ursa Minor	**UMi**	**Ursae Minoris**	**Lesser bear**	**56**	**Polaris (2.0)**
Vela	**Vel**	**Velorum**	**Sails**	**32**	**Gamma Velorum (Suhail) (1.7)**
Virgo	**Vir**	**Virginis**	**Virgin / maiden**	**2**	**Spica (1.0)**
Volans	**Vol**	**Volantis**	**Flying fish**	**76**	**Beta Volantis (3.8)**
Vulpecula	Vul	Vulpeculae	Fox	55	Anser (4.4)

Part One: The Starry Skies

This section will help you get to know the brighter stars and constellations visible throughout the year. Some constellations cover a wide swathe of sky, while others are small enough to be easily obscured by the outstretched hand. Once the names and locations of the brighter stars and constellations are known, they can be used as pointers to less conspicuous celestial objects. This section also contains a taster of some of the brighter deep-sky treasures to be found in selected constellations – there are countless more celestial delights awaiting the eyes of the curious stargazer wishing to take their enjoyment of astronomy to the next level.

Orion and Taurus take centre stage. Imaged using an undriven digital compact camera, 15 second exposure. Overlay of Taurus from Bode's *Uranographia*.

In this section we take a look at the main constellations, stars and celestial showpieces of the northern celestial sphere, beginning with constellations around the north celestial pole and then taking a season-by-season view. Most northern constellations are as familiar to today's stargazers as they were to the ancient Greeks.

Northern circumpolar constellations

From northern temperate climes a number of large constellations – most of them easily identifiable – remain constantly above the horizon throughout the year as they wheel anticlockwise (counterclockwise) around the north celestial pole.

Polaris, the tip of Ursa Minor's tail, is conveniently located just one degree from the north celestial pole. The star is easily found by tracing a line from Merak and Dubhe in Ursa Major, two stars known as the Pointers. These two stars are part of the famous Plough, alternatively known as the Big Dipper or sometimes the Saucepan. One of the sky's most easily identifiable asterisms, the Plough forms the tail and hindquarters of Ursa Major. At midnight during the autumn (fall) the Plough is at its lowest, scraping along the northern horizon, while it soars high overhead at springtime. On the other side of the north celestial pole can be found another prominent asterism, the W of Cassiopeia. The Plough and the W take turns throughout the year in gaining the high celestial ground.

Much of the relatively star-sparse region between the Pointers and the W is filled by the large but ill-defined constellation of Camelopardalis, while another large, faint constellation, Lynx, borders Camelopardalis and Ursa Major. Somewhat easier to trace is the House asterism of Cepheus, which lies between Cassiopeia and the pole. Between Cepheus and the tail of Ursa Major, skirting around the edge of Ursa Minor, is the sprawling constellation of Draco, covering an area of more than a thousand square degrees. Despite being so large, Draco contains only a few reasonably bright stars ranging between the second and third magnitude.

By way of contrast, on the other side of the celestial pole and nestled snugly between Cassiopeia and Auriga, lies beautiful Perseus. For the most part a circumpolar constellation from northern temperate regions, Perseus has a number of bright stars sprinkled along a section of the Milky Way and is one of the loveliest areas to scan with binoculars.

Circumpolar limit from New York (41°N)

Circumpolar limit from London (52°N)

A broad view of the northern circumpolar sky, looking due north (east at right, west at left). The outer circle represents extent of circumpolarity from London (52°N) and the inner circle for stars that are circumpolar from New York (41°N). The ecliptic is also shown at either side (none of it is circumpolar). The chart is relevant for 1 November (4am), 1 December (2am), 1 January (midnight), 1 February (10pm) and 1 March (8pm).

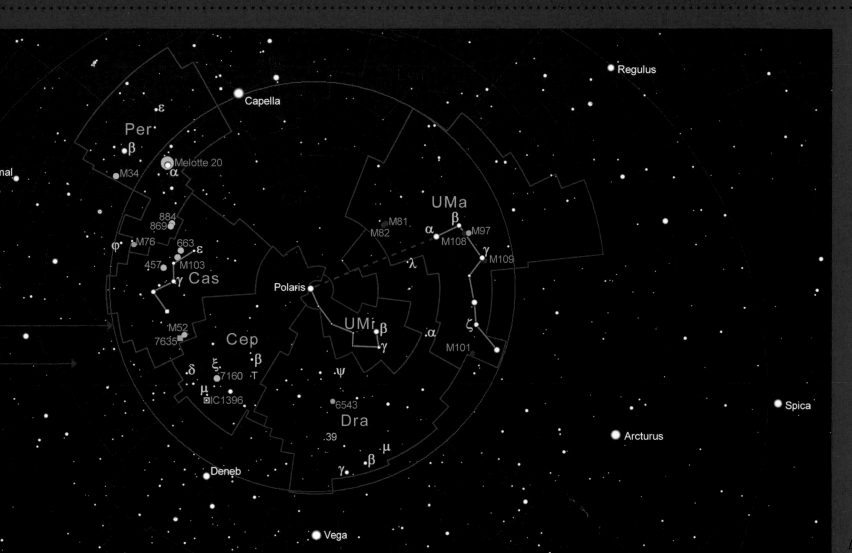

UMI / URSAE MINORIS
Highest at midnight: early January

Sometimes called the Little Dipper, Ursa Minor is a small but significant constellation incorporating the north celestial pole. Its brightest star, **Polaris**, is located within just one degree of the pole. A phenomenon called precession, caused by the slow movement of the Earth's axis, will bring Polaris closest to the pole (within half a degree) at the end of the 21st century.

Polaris is useful to aim at when aligning an equatorial telescope for a quick observing session. A telescope roughly aligned with Polaris will keep an object within the field of view of a medium-power eyepiece for a long time before requiring adjustment. A telescope trained on Polaris itself will reveal it to be a double star, with an eighth-magnitude companion.

    Beta UMi (Kochab) and **Gamma UMi (Pherkad)**, the end stars of the Little Dipper, are sometimes called the Guardians of the Pole. A keen naked eye will discern a faint star close to Pherkad (about half the Moon's diameter away) – this is an unrelated foreground star.

A little circlet of seventh- and eighth-magnitude stars near Polaris can be seen through binoculars – this is nicknamed the 'engagement ring', Polaris being the bright celestial solitaire.

CEP / CEPHEI
Highest at midnight: late February

Cepheus contains several reasonably bright stars. Among the stellar delights of Cepheus is **Beta Cep**, a double star of magnitudes 3.2 and 7.9 that can be resolved through a small telescope. Beta Cep is a variable star, a blue giant whose brightness fluctuates by around one tenth of a magnitude over a period of just a few hours. Another variable, which is also a nice double, is **Delta Cep**. The prototype of the Cepheid variables (*see* Variable stars in Introduction). This is a pulsating yellow supergiant that varies between magnitudes 3.5 and 4.4 in a period of five days nine hours. Delta's companion is a magnitude 6.3 blue star, and the pair is easy to separate through a small telescope.

Mu Cep is a red supergiant whose striking colour has earned it the nickname of the Garnet Star. Binoculars will show its ruddy hue to good effect. The star is also variable, fluctuating between magnitudes 3.4 and 5.1 over a period of two to two and a half years. Mu Cep is one of the biggest stars visible with the unaided eye – if placed in the position of the Sun, the surface of its enormous bloated sphere would extend almost out to the orbit of Saturn.

    Xi Cep is a double star comprising a magnitude 4.4 blue star and a magnitude 6.5 orange giant, easily resolvable through a small telescope.

    T Cep, a red Mira-type variable star, has a period of more than a year. At its maximum

The Garnet Star in Cepheus, observed through a 100mm refractor by the author.

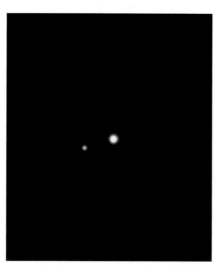

Double star Delta Cephei.

brightness the star just pops into the range of the unaided eye, shining at magnitude 5.2.

An extension of the Milky Way nudges into the southern part of Cepheus, and a couple of lovely open star clusters can be found in this vicinity – **IC 1396** and **NGC 7160** – both delightful to view through a 150mm telescope. IC 1396 is embedded within a sizeable patch of nebulosity; it is also known as the **Elephant's Trunk Nebula** because of a prominent dark sinuous dust lane, which is visible on photographs. Large binoculars will reveal IC 1396 as a misty patch. NGC 7160 is a small, compact star cluster; around 30 of its stars are visible through a 200mm telescope, half a dozen of the brighter ones standing out from the rest.

The Elephant's Trunk Nebula in Cepheus, imaged through an 80mm refractor with an astronomical CCD camera (filters used).

UMA / URSAE MAJORIS
Highest at midnight: early March

Seven of the brightest stars within Ursa Major make up an asterism variously called the Plough or the Big Dipper. While this asterism itself doesn't much look like a bear, a little time spent in tracing the traditional outline comprising the remainder of the constellation's bright stars will convince any stargazer that the ancients who named it had an extremely good eye for form.

The two front stars of the Plough, **Alpha UMa (Dubhe)** and **Beta UMa (Merak)** are known as the Pointers, since an imaginary line extending from them leads to Polaris and the north celestial pole. **Zeta UMa (Mizar)**, the second star of the Plough's handle, has a fainter magnitude 4 partner, **80 UMa (Alcor)**, which is visible with the unaided eye. Mizar itself is a close double star, with components of magnitudes 2.2 and 3.8, separable with a small telescope.

A pair of galaxies bright enough to be seen through binoculars, **Bode's Galaxy (M81)** and the **Cigar Galaxy (M82)** lie in the far north. Just half a degree apart, the pair is visible in the same low-power telescopic field. While M82 is almost edge-on to us, M81 is tilted at less of an angle. Some ten million light years distant, these galaxies are interacting with each other. On the other side of the constellation, the face-on spiral galaxy **M101** is visible through binoculars as a circular smudge, and appears mottled through a 200mm telescope.

The **Owl Nebula (M97)**, a faint planetary nebula, appears as a pale disk about twice the diameter of Jupiter through a 150mm telescope. The dark eyes of the owl, so obvious in many images, are

The familiar stars of the Plough in Ursa Major.

Multiple star Mizar in Ursa Major.

rather elusive and require at least a 250mm telescope to discern. **M108**, a bright, sizeable and nearly edge-on galaxy, can fit into the same low-power telescopic field as M97. Bright condensations within M108 can be discerned through a 150mm telescope.

Although the location of the galaxy **M109** can be identified fairly easily, being a little more than half a degree east of **Gamma UMa**, its low surface brightness makes it more of a challenge to observe. A 200mm telescope will reveal its bright elliptical centre along with a superimposed foreground star just to the north of the core, but detail within the spiral arms requires a larger instrument to resolve.

Galaxy M101 in Ursa Major, imaged using a 127mm refractor and astronomical CCD camera.

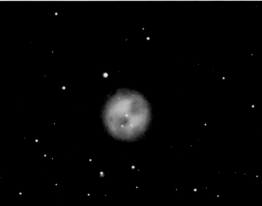

Planetary Nebula M97 in Ursa Major, imaged using a 127mm refractor and astronomical CCD camera.

♌ DRA / DRACONIS
Highest at midnight: early July

Despite sprawling across a huge portion of the northern circumpolar region, Draco is not the most prominent of constellations. Its traditional outline can be traced from its head (marked by **Beta Dra** and **Gamma Dra**) just north of Hercules, along a winding path to **Alpha Dra (Thuban)** at the constellation's narrowest part, around to **Lambda Dra** near Draco's western border. Around the time that the Pyramids were constructed, Thuban (magnitude 3.7) was the brightest star near the north celestial pole; precession will grant it this distinction again in more than 21,000 years' time.

Mu Dra is a close telescopic double star with white components of magnitudes 4.9 and 5.6; the pair is slowly moving apart, and they can now be comfortably resolved with a 100mm telescope, and are a good test for a 60mm telescope. **Psi Dra** is much easier to resolve; binoculars will reveal the yellow stellar duo of magnitudes 4.6 and 5.8. Binoculars can split the two wide components of **39 Dra** (magnitudes 5 and 7.4); a telescope will show the magnitude 8 companion of the brighter star.

The Cat's Eye Nebula (NGC 6543) is a small but bright planetary nebula with a distinct bluish hue. A 150mm telescope will show it as a small ring surrounding an eleventh-magnitude central star.

The Cat's Eye Nebula in Draco, imaged using a 105mm refractor and astronomical CCD camera.

CAS / CASSIOPEIAE
Highest at midnight: early October

With its prominent five-star W asterism, Cassiopeia is one of the easiest constellations to recognize. **Gamma Cas**, the central star of the W asterism, is an irregular variable star that fluctuates, at unpredictable intervals, between magnitudes 3 and 1.6. **Eta Cas** is a nice double star with a magnitude 3.5 yellow primary and a red magnitude 7.5 companion, easily visible through a small telescope.

Cassiopeia is a joy to scan with binoculars, as a bright section of the Milky Way flows across the constellation, engulfing the W. A treasure trove of bright open clusters lies within its boundaries, most of which lie east of the W. Containing around 30 stars, the bright compact cluster of **M103** is best seen at higher magnifications. The **Owl Cluster (NGC 457)** is a loose assembly of around 100 fairly bright stars arranged in distinct lines; its two brightest stars shine like an owl's eyes. **NGC 663** is a beautiful binocular cluster containing around 80 stars. On the far western side of Cassiopeia, the compact **Scorpion Cluster (M52)** contains around 100 stars, the brightest of which form a splendid S shape. The **Bubble Nebula (NGC 7635)**, a faint diffuse nebula visible through a 200mm telescope, lies just half a degree southwest of M52, so that the two objects can be viewed in the same low-power field of view.

A delightful alignment of stars known as Kemble's Cascade can be found in the circumpolar constellation of Camelopardalis, east of Cassiopeia.

The Bubble Nebula in Cassiopeia, imaged using an 80mm refractor and astronomical CCD camera (filters used).

Perseus

PER / PERSEI
Highest at midnight: mid-November

Crossed in the north by the Milky Way, Perseus is a magnificent constellation containing a number of bright stars and open clusters. Near **Alpha Per (Mirfak)** lies **Melotte 20**, a large loose star cluster made up of a snaking chain of bright stars; it is a stunning sight through binoculars and at low magnifications.

Beta Per (Algol) is a famous eclipsing binary. Every 2.87 days it drops from magnitude 2.1 to 3.4, changes easily monitored with the unaided eye. **The Spiral Cluster (M34)** can just be discerned with the unaided eye some five degrees west of Algol. It contains a number of star chains, with some of its brighter stars paired up.

Eta Per is a nicely coloured double star, easily resolvable through a small telescope, with an orange magnitude 3.8 primary and a blue magnitude 8.5 companion.

Located in the far northwestern corner of Perseus, the **Double Cluster (NGC 869 and NGC 884)** is one of the most breathtaking sights in the heavens. These two bright open clusters – each the diameter of the full Moon – lie side by side, and can be glimpsed as a hazy patch with the unaided eye. A low-power view will accommodate both clusters, revealing hundreds of stars. Several red stars can be discerned near NGC 884, contrasting nicely with the cluster's profusion of blue stars.

Glowing at magnitude 10, the **Little Dumbbell Nebula (M76)** is the faintest Messier object (*see* Telescopic revelations in Introduction). Resembling a tiny apple core, this faint planetary nebula can be seen through a 150mm telescope. It lies less than one degree north of **Phi Per**.

Planetary nebula M76 in Perseus, imaged using a 160mm refractor and astronomical CCD camera (filters used).

Northern winter stars (midnight, 1 January) ....

Plenty of dazzling stars and bright deep-sky objects are sprinkled about the winter night skies. The Milky Way, dotted with fabulous star clusters, runs from near the zenith to the southern horizon, while the ecliptic slices across the skies, from Virgo in the east, through Leo, Cancer and Gemini to Taurus, Aries and Pisces in the west.

Orion, climbing high above the southern horizon is immediately recognizable, with its bright orange star Betelgeuse, dazzling Rigel and the three bright stars in a line making up Orion's Belt. Below, in Orion's Sword Handle, can be seen a fuzzy patch – the glow of the Orion Nebula, one of the most beautiful of all deep-sky objects.

Orion's Belt stars can be used as a handy signpost for finding other stars and constellations. Follow the belt downwards to the left and you'll come to Canis Major and Sirius, the brightest star in the night skies. Because of its relatively low altitude from mid-northern latitudes, Sirius often appears to twinkle because of the effects of Earth's atmosphere, producing a multicoloured display of stellar scintillation.

Extend the line of Orion's Belt to the right and you'll find Aldebaran, a bright orange star in Taurus. The V-shaped asterism to which Aldebaran belongs, represented as the horns of a bull on old star charts, is called the Hyades; further to the west lies the Pleiades, a cluster of bright stars.

Above Orion lies Auriga, its brightest star Capella soaring almost directly overhead.

Straddling the Milky Way, Auriga contains a number of beautiful bright open star clusters. Neighbouring Auriga, further along the band of the Milky Way, lies the constellation of Perseus. To its west lie two of the largest galaxies in our Local Group – M33, the Pinwheel Galaxy and M31, the Great Andromeda Spiral. Both of these nearby galaxies are comparable in size to our own and can be glimpsed with the unaided eye under very dark skies.

High in the south climb the heavenly twins Castor and Pollux in Gemini whose toes dip into the Milky Way. At its heels follows Canis Minor and the bright star Procyon. Meanwhile, as Virgo rises in the east it is preceded by Leo, with its familiar Sickle asterism and bright star Regulus, and the fainter constellation of Cancer. This contains a large open cluster, the Beehive, easily discerned with the unaided eye between Regulus and Pollux.

Looking towards the northern horizon, the familiar circumpolar stars have completed another quarter-turn about the celestial pole. Ursa Major is in the ascendancy standing on its tail, while Cassiopeia is beginning to sink westwards.

● Arcturus

Horizon from New York (41°N) ⟶

Horizon from London (52°N) ⟶

● Spica

Northern winter sky, looking due south (east at left, west at right) from the horizon to the zenith. The horizon lines for London (52°N) and New York (41°N) are marked, as well as the ecliptic. The chart is relevant for 1 November (4am), 1 December (2am), 1 January (midnight), 1 February (10pm) and 1 March (8pm).

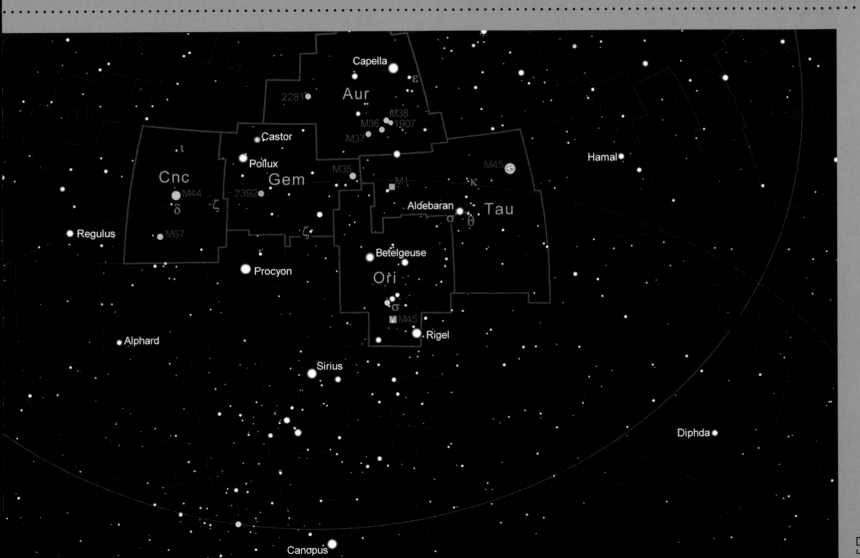

Taurus ....

TAU / TAURI
Highest at midnight: early December

Brooded over by the red bulls' eye of **Alpha Tau (Aldebaran)**, this large and easily identifiable Zodiacal constellation dominates the skies northwest of Orion. Aldebaran is set against the background of the **Hyades**, a large V-shaped open cluster containing a dozen or more naked-eye stars in an area that can be covered by a clenched fist. Among them, the bright doubles **Theta Tau** and **Kappa Tau** are both separable with a keen unaided eye, but **Sigma Tau** requires some optical aid to split.

In the northwest of Taurus, a handful of the brightest stars within the **Pleiades (M45)** star cluster is easy to see with the unaided eye. Telescopes will reveal many dozens of young blue stars, and near the Pleiad **23 Tau (Merope)** may be seen a hint of reflecting nebulosity.

Just over one degree north of **Zeta Tau**, the faintly glowing supernova remnant of the **Crab Nebula (M1)** requires an 80mm telescope to be seen at all well. Even through large instruments it appears as a grey, rather featureless elliptical patch.

Wispy nebulosity can be seen surrounding the Pleiades star cluster, imaged with a 105mm refractor and astronomical CCD camera.

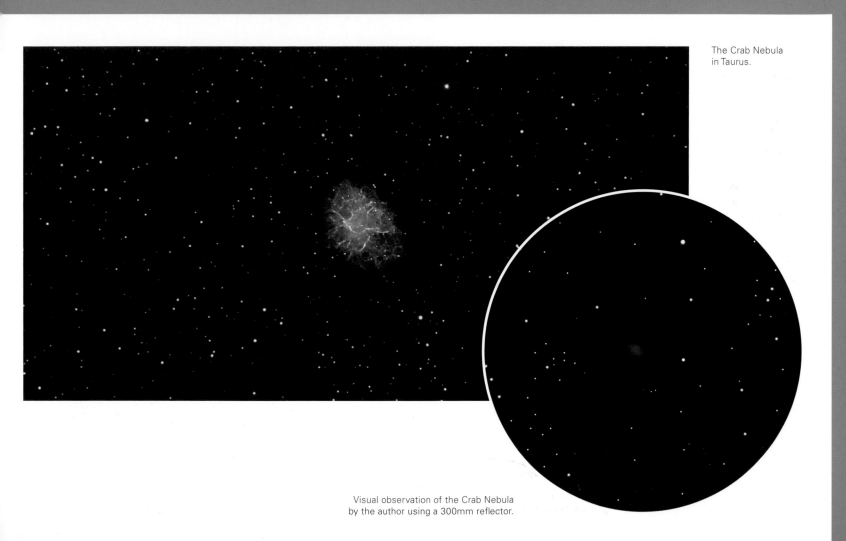

The Crab Nebula
in Taurus.

Visual observation of the Crab Nebula
by the author using a 300mm reflector.

Orion

ORI / ORIONIS
Highest at midnight: mid-December

Orion is the most magnificent of all the constellations. Straddling the celestial equator, its main shape – made up of the red **Alpha Ori (Betelgeuse)** in the north, the brilliant blue **Beta Ori (Rigel)** in the south and the three dazzling stars of **Orion's Belt** in between – is recognizable to stargazers in both northern and southern hemispheres.

Sigma Ori is a lovely multiple; its main star of magnitude 3.8 has two nearby sixth-magnitude stars, and a more distant triple of eighth-magnitude stars. All six stars can be marvelled at in a high-power field of view.

A short distance south of Orion's Belt, a misty patch can be discerned with the keen unaided eye. This is the **Orion Nebula (M42)**, one of the biggest and brightest nebulae in the heavens. Considerable structure within the nebula can be seen with binoculars alone, and a small telescope will reveal a glowing greenish mass with delicate wisps, intruded on by a prominent dark lane. Several stars can be seen in and around the nebula, notably the **Trapezium (Theta Ori)**, a bright quadruple star. Larger telescopes will show breathtaking detail within the nebula. Just to its north lies the fainter **De Mairan's Nebula (M43)**, with a single bright star nestling at its centre.

The beautiful Horsehead Nebula in Orion is very challenging to view through even large amateur telescopes, but it is captured here on a CCD image taken through a 127mm refractor (filters used).

Auriga

AUR / AURIGAE
Highest at midnight: mid-December

Broad, bright and easily recognizable, Auriga rides high overhead on winter nights. Southwest of brilliant **Alpha Aur (Capella)** lies the small triangular asterism of the **Kids**. Its apex star, **Epsilon Aur**, is a white supergiant of magnitude 3 that is eclipsed by an unseen companion every 27 years, when it drops to magnitude 3.8 for a year. Its next eclipse is due to take place around 2037.

Three of the brightest clusters in Auriga – **M36**, **M37** and **M38** – are easily visible through binoculars as misty patches in the southern part of the constellation. Of these, M37 is the biggest and brightest; an orange ninth-magnitude star lies at the centre of around 150 fainter stars, resolvable with a 150mm telescope, with a hint of nebulosity at its centre. M38 is made up of around 100 stars below tenth magnitude, the brightest of which form a startling cross shape. Telescopes will reveal the much smaller and fainter cluster NGC 1907 half a degree south of M38. In eastern Auriga, the cluster NGC 2281 contains a wide scattering of half a dozen fairly bright stars, with more than 20 fainter ones in the background.

Star cluster M36 in Auriga, imaged with a 160mm refractor and astronomical CCD camera (filters used).

Gemini

GEM / GEMINORUM
Highest at midnight: early January

Castor, the brightest star in Gemini.

A sizeable, easily recognized constellation, Gemini is a bright and well-defined constellation lying between Auriga and Canis Minor. Its two main stars, **Alpha Gem (Castor)** and **Beta Gem (Pollux)** are instantly identifiable. Pollux, slightly the brighter of the pair, has a decidedly orange hue. Castor is a famous multiple star, with its two brightest components (magnitudes 1.9 and 3) resolvable through a good 60mm telescope. Interestingly, both **Eta Gem** and **Zeta Gem** are double stars whose primaries are also variables.

The far western part of Gemini is immersed in the Milky Way. Here, the beautiful open cluster **M35** can be glimpsed with the unaided eye, just a couple of degrees northwest of Eta Gem – binoculars show it well. Spread across an area the size of the full Moon, M35 contains about 80 stars, some of which are arranged into a prominent curved chain.

A couple of degrees southeast of **Delta Gem** can be found the **Eskimo Nebula (NGC 2392)**, a bright eighth-magnitude planetary nebula that shows up as a greenish blob at low magnifications. It is surprisingly large as planetary nebulae go, with a diameter equivalent to the apparent size of Jupiter. It has a bright, almost stellar centre.

Star cluster M35 in Gemini, imaged with a 160mm refractor and astronomical CCD camera (filters used).

Cancer

CNC / CANCRI
Highest at midnight: early February

Cancer is the least prominent of all the 12 Zodiacal constellations. Although it is so faint, the constellation can easily be located to the southeast of the bright stellar twins Castor and Pollux. Cancer's third- and fourth-magnitude brightest stars form the shape of an inverted Y, which is easy to trace from dark-sky sites.

Small telescopes will split the two main components of **Zeta Cnc**, widely separated stars of magnitudes 5.2 and 5.8 and a 200mm telescope will reveal that the brighter star has a closer magnitude 6.2 companion. **Iota Cnc** is a lovely coloured double, with a yellow magnitude 4 primary and a blue magnitude 6.6 companion, easy to resolve with a small telescope.

Keen eyes will discern a misty patch just north of **Delta Cnc** and directly in the middle of the rectangular area bounding the constellation. Binoculars will reveal this to be the **Beehive Cluster (M44)**, a sizeable star swarm covering an area of around ten times that of the full Moon. A very low telescopic magnification will take in the entire cluster. Around 80 stars can be seen, of which the sixth-magnitude **Epsilon Cnc** is the brightest.

Less than two degrees west of **Alpha CnC** lies a much smaller open cluster than the Beehive, the **King Cobra (M67)**. Despite its large near neighbour, M67 is a major open cluster in its own right, containing 300 stars and appearing as a full Moon-sized oval smudge through binoculars. A 150mm telescope will resolve the cluster's fainter stars, most of which are below eleventh magnitude.

Star cluster M67 (the King Cobra) in Cancer, imaged with a 105mm refractor and astronomical CCD camera.

Northern spring stars (midnight, 1 April) ....

With its milder nights, springtime is eagerly looked forward to by those who prefer milder observing conditions. Having revolved to its highest point in the sky, Ursa Major and the Plough soar high overhead. Cassiopeia is low above the northern horizon, its W asterism perhaps a little difficult to discern from light-polluted urban areas. More of a challenge to view with the unaided eye, the band of the Milky Way runs at its lowest across the sky, almost parallel to the northern horizon, from Canis Minor in the west through Perseus and Cassiopeia in the north to Aquila in the east. Procyon, Canis Minor's brightest star, is becoming increasingly low in the west and often appears to twinkle because of its low altitude.

Having ascended to near-zenithal heights during the winter months, Capella, the brightest star in Auriga, now slowly sinks towards the northwestern horizon, followed by Castor and Pollux in Gemini. Hard on their heels Leo, with its bright star Regulus, prowls above the southwestern horizon, tentatively eyeing the faint constellation of Cancer. High in the south, trailing Leo, is the faint constellation of Coma Berenices, notable for its broad cluster of dim stars – the cluster appears as a faint patch of light just beyond most people's naked-eye resolution and is sometimes mistaken for a cloud.

Both Coma Berenices and Virgo to its south contain the richest collection of bright galaxies in the entire sky. Deep-sky observers eagerly explore the Virgo-Coma Cluster – a grouping of more than 1,300 galaxies of varying sizes and shapes lying around 50 million light years away – some of which are bright enough to be seen through binoculars and small telescopes.

Fainter constellations nearby include Libra, Corvus, Crater and Sextans, all underlain by lengthy Hydra whose brightest star, Alphard, lies midway between Regulus and the southwestern horizon. Stretching across the sky from Libra in the southeast, passing Spica and Regulus, the ecliptic reaches to Taurus, low in the northwest.

On the ascendancy, brilliant Lyra in Vega is climbing above the northeastern horizon, followed by Deneb in Cygnus. Orange Arcturus in Boötes shines brightly in the southeast, while Hercules and the pretty constellation of Corona Borealis are climbing to an increasingly favourable altitude.

Horizon from New York (41°N) ● Altair

Horizon from London (52°N)

Northern spring sky, looking due south (east at left, west at right) from the horizon to the zenith. The horizon lines for London (52°N) and New York (41°N) are marked, as well as the ecliptic. The chart is relevant for 1 February (4am), 1 March (2am), 1 April (midnight), 1 May (10pm) and 1 June (8pm).

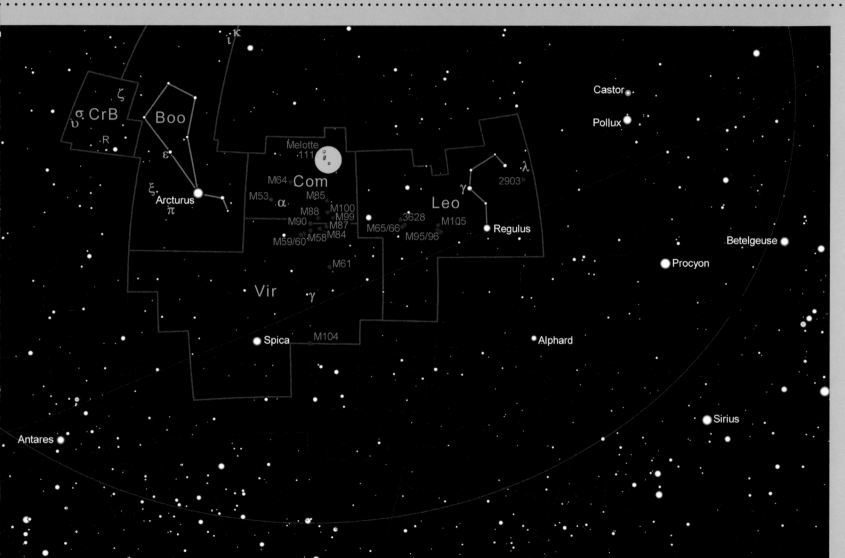

ι κ

σ CrB
υ
ζ
R

Boo
ε
ξ
Arcturus
π

Melotte
111

M64
M53
M88
M90
M59/60
M58
M84
M87
M99
M100
M85
α
Com

3628
M65/66
M95/96
M105

Leo
γ
λ
2903

Castor
Pollux
Regulus

Betelgeuse

Procyon

M61

Vir
γ

Alphard

Spica
M104

Sirius

Antares

Leo

LEO / LEONIS
Highest at midnight: early March

Leo dominates its immediate neighborhood. It is easy to find, with the prominent Sickle asterism in the west and a bright triangle of stars forming its tail in the east.

Alpha Leo (Regulus) at the base of the Sickle is a bright double star whose components of magnitudes 1.4 and 7.7 can be easily separated through a small telescope. Being located less than half a degree from the ecliptic, Regulus is often occulted by the Moon. **Gamma Leo** is another charming double, made up of a pair of yellow stars of magnitudes 2.3 and 3.6.

A number of bright galaxies lie in the belly of Leo. **M65** and **M66**, a pair of ninth-magnitude spiral galaxies, are less than half a degree apart; just to their north lies a fainter edge-on galaxy, **NGC 3628**. All three can be viewed in a low-power field through a 100mm telescope. Another notable triplet of galaxies lies a short distance to the west and comprises **M95**, **M96** and **M105**. M95 is face-on with a bright nucleus, while M96 is a rather homogenous round smudge devoid of a nucleus. M105 is a cigar-shaped elliptical blur. The easily located spiral **NGC 2903** lies one and a half degrees south of **Lambda Leo**, and shows some blotchiness through a small telescope.

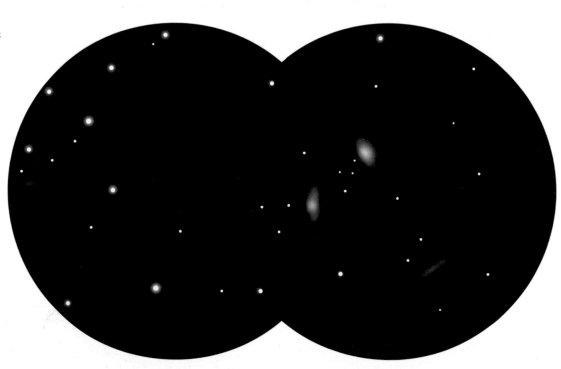

Galaxies M65 and M66, observed by the author using a 300mm reflector (two telescopic fields of view combined); also visible are the fainter galaxies NGC 3628 (lower right) and NGC 3593 (far left).

COM / COMAE BERENICES
Highest at midnight: early April

Coma is an inconspicuous little constellation made up of around 30 faint stars on the border of naked-eye visibility. The eye's attention is drawn to an extensive misty patch, the **Coma Star Cluster (Melotte 111)** containing around 40 stars (mostly below naked-eye brightness) spread over a five-degree wide area.

The Black Eye Galaxy in Coma, imaged with a 127mm refractor and astronomical CCD camera.

Hundreds of stars within **M53**, a bright globular cluster with a concentrated core, can be viewed through a 200mm telescope at high magnification. The cluster is easy to locate, just one degree northeast of **Alpha Com**.

All the brighter galaxies of Coma lie within the **Coma-Virgo Cluster**, a vast assembly of galaxies around 50 million light years away. Probably the most spectacular is the **Black Eye Galaxy (M64)**, a bright galaxy whose prominent silhouetted dark lane is easily visible through a 150mm telescope. There are many more Coma galaxies easily visible through small telescopes, including **M85**, **M88**, **M99** and **M100**.

Virgo ....

♍ VIR / VIRGINIS
Highest at midnight: mid-April

Stretching large and broad along the celestial equator, Virgo, the sky's second biggest constellation, occupies almost 1,300 square degrees of sky. Once the bright **Alpha Vir (Spica)** is located, the other stars making up its pattern can be traced fairly easily. To give the stargazer an idea of scale, two outstretched hand widths, with Spica beneath the intersection of the thumbs, will just about cover the constellation's main body.

Gamma Vir (Porrima) is a famous double star of equal components, both white stars shining at magnitude 4.6. The pair orbit one another in a period of around 170 years, and were closest in 2008 when they were only separable at high magnification using a 250mm telescope. In 2020 a good 60mm telescope will be able to resolve them, and they will be at their widest in 2080, when any small telescope will split them.

All of the bright galaxies in Virgo belong to the **Coma-Virgo Cluster**. Most of them lie in the northwest of the constellation, and include **M58**, **M59**, **M60**, the **Swelling Spiral (M61)**, **M84**, **M87** and **M90**. In the southwest of Virgo can be found the **Sombrero (M104)**, a bright eighth-magnitude edge-on galaxy whose nuclear bulge rises smoothly on either side of its spindly spiral arms. A dark dust lane running through the centre of M104 is visible through a 300mm telescope.

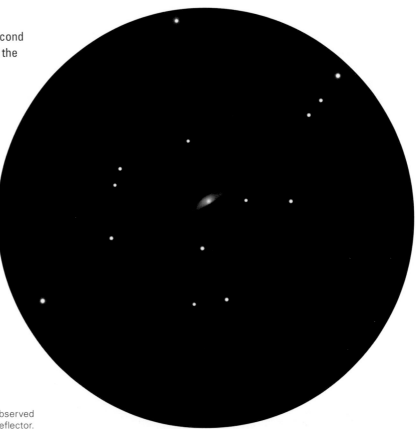

The Sombrero Galaxy in Virgo, observed
by the author with a 300mm reflector.

BOO / BOÖTIS
Highest at midnight: early May

CRB / CORONAE BOREALIS
Highest at midnight: late May

The brightest star in Boötes, the orange giant **Alpha Boo (Arcturus)** can be located by tracing a slightly curving line from the tail of Ursa Major. Many stargazers trace out the shape of Boötes by imagining a large kite, with Arcturus at its tail end and **Beta Boo** at its apex; the image is completed by a trail of stars west of Arcturus, representing the kite's trailing ribbon.

Although Boötes is a large constellation, it contains no bright deep-sky objects. It makes up for this with a number of fine double stars. **Epsilon Boo** is a lovely close double with an orange magnitude 2.5 primary and a blue magnitude 4.6 companion, and requires a 100mm telescope at x100 to resolve well. Doubles that are easily visible through a 60mm telescope include **Iota Boo**, a wide double of magnitudes 4.8 and 8.3; **Kappa Boo**, an easy double of magnitudes 4.5 and 6.6; **Pi Boo**, magnitudes 4.5 and 5.8; and **Xi Boo**, a beautiful yellow-orange double of magnitudes 4.7 and 7.

A small but delightful constellation, the main pattern of Corona Borealis is made up of a semicircle of seven bright stars.

Several doubles are easily separable through small telescopes, including **Zeta CrB** (magnitudes 5 and 6), **Nu CrB** (a wide fifth-magnitude pair) and **Sigma CrB** (magnitudes 5.6 and 6.6). **R CrB** is a dwarf nova cataclysmic variable star whose minima occur randomly. Mostly it shines around magnitude 6, but it can suddenly dim by up to eight magnitudes and can remain faint for many weeks.

Arcturus (Greek, meaning 'bear keeper'), the sky's fourth brightest star.

Boötes and Corona Borealis

Northern summer stars (midnight, 1 July) ....

While gaining in one hand, the night sky observer loses in the other, as shirtsleeves observing conditions are accompanied by shorter, lighter nights. All the familiar winter constellations are below the horizon, although it is still possible to glimpse Capella in Auriga, a circumpolar star, low above the northern horizon.

Ursa Major has rotated around to the northwest, the Saucepan poised as if to pour its cosmic contents on faint Lynx below. Following the curve of the bear's tail takes us to Arcturus, which is slowly dropping in the western skies. The galactic feast begun in spring is coming to an end as Coma Berenices and Virgo sink into the west, only to be replaced with a sumptuous sampling of our own home Galaxy, the Milky Way.

Hercules, Corona Borealis and Ophiuchus – all lovely constellations with a wonderful selection of bright celestial gems – precede the famous Summer Triangle asterism made up of Vega in Lyra (now near the zenith), Deneb in Cygnus and Altair in Aquila. From the low southern constellations of Sagittarius and Scorpius, arching high through the Summer Triangle and into Cassiopeia and Perseus, can be seen the most spectacular section of the Milky Way visible from the northern hemisphere. When viewed from a dark-sky site, unhindered by light pollution, the unaided eye can discern remarkable detail along this section of the Milky Way. A prominent dark rift, caused by interstellar gas and dust seen in silhouette against distant stars in our

Galaxy, runs through the Northern Cross asterism in Cygnus; further south, a seemingly detached section of the Galaxy known as the Scutum Star Cloud gives the northern hemisphere observer some idea of what the Magellanic Clouds look like to those living in the southern hemisphere.

Andromeda and the Square of Pegasus asterism, floating on an area of sky known to the ancients as the Water – containing the constellations of Pisces, Aquarius and Capricornus the Sea Goat – rise in the east. Cutting its lowest path across the skies for northern hemisphere observers, the ecliptic runs from faint Pisces in the southeast, through Sagittarius, low in the south, to Virgo in the west; at its highest the ecliptic is barely a hand's width above the southern horizon. Our Galaxy's central hub, located in Sagittarius, is masked by dark interstellar clouds viewed in silhouette, while a good sprinkling of beautiful bright nebulae can be seen in the vicinity.

Hamal

Horizon from New York [41°N]

Horizon from London [52°N]

Northern summer sky, looking due south (east at left, west at right) from the horizon to the zenith. The horizon lines for London (52°N) and New York (41°N) are marked, as well as the ecliptic. The chart is relevant for 1 May (4am), 1 June (2am), 1 July (midnight), 1 August (10pm) and 1 September (8pm).

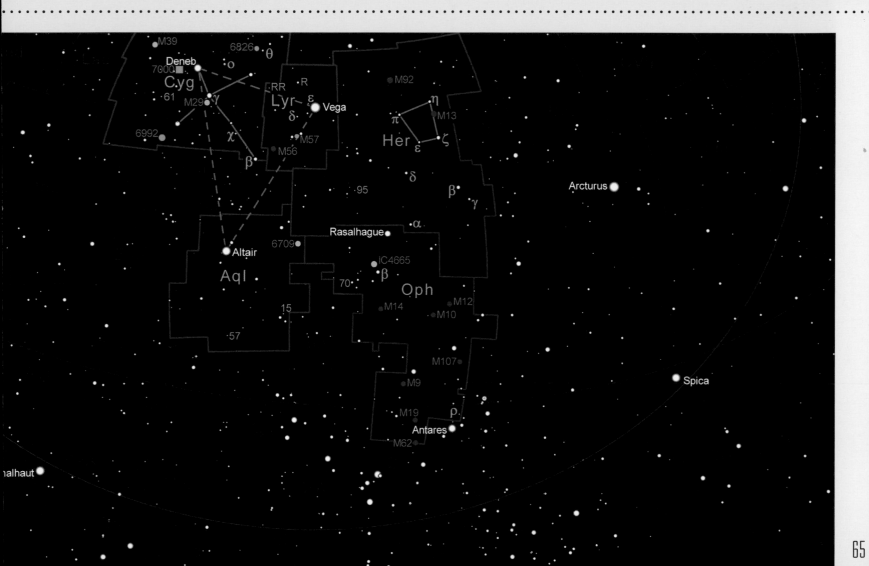

Hercules

The Great Globular Cluster M13 in Hercules, observed by the author with a 300mm reflector.

♓ HER / HERCULIS
Highest at midnight: early June

A large constellation, identifiable by the **Keystone** asterism made up of four of Hercules' brighter stars – **Pi Her**, **Eta Her**, **Epsilon Her** and **Zeta Her**. These stars, combined with **Beta Her** and **Delta Her** further south, make up a familiar big butterfly-shaped asterism. The constellation stretches considerably further north, south and east of this, making Hercules the sky's fifth largest constellation.

Alpha Her is a splendid red giant varying between magnitudes 3 and 4; a small telescope will show that it has a close green-coloured companion of magnitude 5.4. **Gamma Her** and **95 Her** are also nice telescopic doubles.

The **Great Globular Cluster (M13)** is the northern sky's brightest example of its type. Lying just two and a half degrees south of Eta Her, M13 is easy to locate – indeed, it is faintly visible with the unaided eye. Binoculars show it to be an extensive fuzzy patch measuring around half the apparent diameter of the full Moon. Viewed through a 150mm or larger telescope, the cluster is an amazing sight; the brightest of its 300,000 outlying stars can be resolved, and these appear to be arranged in several distinct radial lines. Hints of darker lanes can be discerned within the cluster's outer regions; photographs don't show these features well, but our perception through the eyepiece produces a different impression.

M92 is probably the northern sky's second most beautiful globular cluster. Located north of the Keystone, it receives less attention than its brighter sibling, but it is in many ways just as spectacular. Its outer stars can be resolved through a 150mm telescope, and the cluster is smaller and more compact and spherical than those of M13.

Globular cluster M92 in Hercules. CCD image taken with a 160mm refractor.

Ophiuchus ....

OPH / OPHIUCHI
Highest at midnight: mid-June

A large constellation that extends well south of the celestial equator, Ophiuchus's main stars are of the second and third magnitude. Its full outline can be hard to trace from northern climes, since it extends to 30 degrees south.

The multiple star **Rho Oph** makes a great high-magnification sight. It consists of a magnitude 4.6 primary and close magnitude 5.7 partner, plus two more widely separated outlying stars of seventh magnitude. Another double worthy of note, **70 Oph**, comprises components of magnitudes 4.2 and 6, which orbit each other in a period of 88 years; the pair will be at its widest in 2025.

North of **Beta Oph** lies **IC 4665**, a scattered open cluster visible with the naked eye as a misty patch, best viewed through binoculars. Several degrees to the east lies **Barnard's Star**, a ninth-magnitude red dwarf, under six light years distant.

Ophiuchus is rich in globular clusters, seven of the brightest appearing in Messier's list (*see* Introduction): **M9**, **M10**, **M12**, **M14**, **M19**, **M62** and **M107**. The brightest of these, M10 and M12, can be resolved through a 200mm telescope.

Globular cluster M14 in Ophiuchus, imaged with a 127mm refractor and astronomical CCD camera.

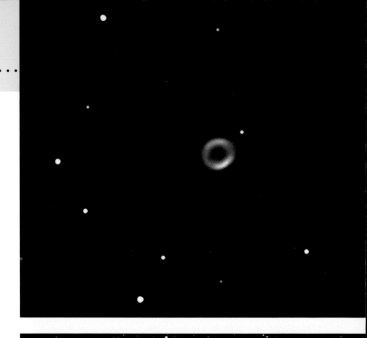

Lyra ....

LYR / LYRAE
Highest at midnight: early July

The Ring Nebula in Lyra, imaged with a 200mm SCT and astronomical CCD camera.

The compact constellation of Lyra is one of the best-known constellations of northern summer skies. Its western margin is clipped by a rich, wide section of the Milky Way. **Alpha Lyr (Vega)** is the fifth brightest star in the sky, and the second brightest in the northern celestial hemisphere. Vega is 25 light years away, measures more than three times the diameter of the Sun and is 61 times brighter. Vega was the first star ever photographed, back in July 1850. In 12,000 years' time, precession of the Earth's axis will have moved the north celestial pole near Vega.

Beta Lyr is a beautiful double star, the primary being a white variable star of magnitude 3.3–4.4; its companion a blue magnitude 7.2, **Delta Lyr**. The **Double-Double Star (Epsilon Lyr)** is a famous multiple. The main pair is wide enough to be separated with binoculars, while each of these is a close telescopic double of magnitudes 4.6 and 5.3, and magnitudes 4.7 and 6.1, all comfortably resolvable through a 100mm telescope.

  R Lyr, a variable star, is a red giant that pulsates between magnitudes 3.9 and 5 every six to seven weeks. Some distance to its east lies **RR Lyr**, a special type of pulsating variable, which varies between magnitudes 7.1 to 8.1 in 13.6 hours.

  The **Ring Nebula (M57)** is perhaps the best-known planetary nebula. Easily found midway between Beta Lyr and Gamma Lyr, M57 is rather small; binoculars show it as an almost star-like point, but through a telescope at a high magnification it resembles a sharply defined luminous ring. The brightest of Lyra's other deep-sky delights is the globular cluster **M56**. Many of its stars are resolvable through a 200mm telescope, and are set in a lovely rich galactic starfield.

Globular cluster M56 in Lyra, imaged with a 160mm refractor and astronomical CCD camera.

Aquila

AQL / AQUILAE
Highest at midnight: mid-July

Aquila, a medium-sized constellation, straddles the celestial equator. **Alpha Aql (Altair)**, the most southerly star of the prominent **Summer Triangle**, is one of our nearest stellar neighbours, lying just 17 light years away. Aquila contains two nicely coloured double stars, easily visible through small telescopes: **15 Aql**, a magnitude 5.4 orange star with a lilac magnitude 7 companion; and **57 Aql**, a sky-blue magnitude 5.7 primary with a magnitude 6.5 companion.

Aquila contains a fair scattering of faint planetary nebulae, but its deep-sky showpiece is **NGC 6709**, an open cluster made up of around 30 fairly bright stars, a number of which are arranged in loose chains.

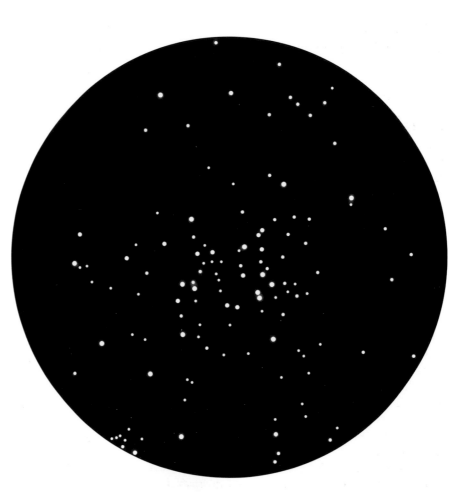

Star cluster NGC 6709 in Aquila, observed by the author with a 100mm refractor.

Cygnus ....

2 CYG / CYGNI
Highest at midnight: early August

The North America Nebula in Cygnus, imaged with an astronomical CCD camera.

Flying south along the Milky Way, the **Swan** is a wonderful constellation set against the bright, rich starry background of the Milky Way. The stargazer can spend literally hours scanning this area through binoculars, sweeping along magnificent starfields and spending a while searching for some of the more elusive deep-sky quarry on offer. As well as being one of the three bright stars that make up the Summer Triangle asterism, **Alpha Cyg (Deneb)** is the top star of the **Northern Cross** asterism. At the foot of the cross, near the southwestern border of Cygnus, is **Beta Cyg (Albireo)**, one of the most beautiful coloured double stars. A small telescope will easily resolve the companion to its golden magnitude 3.1 primary, a steely blue star of magnitude 5.1. **Omicron Cyg** is another wonderful coloured double, separable through binoculars, made up of an orange magnitude 3.8 primary and a sea-green magnitude 4.8 companion. Closer scrutiny will reveal another companion to the primary, a blue magnitude 7 star.

The red giant **Chi Cyg**, a Mira-type variable star, can reach the third magnitude at its brightest, although it normally shines well below naked-eye visibility, dropping to as low as the fourteenth magnitude. Its period is around 400 days. **61 Cyg** is a double star made up of two orange dwarfs, magnitudes 5.2 and 6, easy to split through a small telescope.

M29, a loose open cluster of around 20 stars, of which several are reasonably bright, is easy to find by sweeping south of **Gamma Cyg**. M39, a much bigger open cluster, is

Double star Albireo, a beautiful contrasting pair of blue and gold set amid the Milky Way in Cygnus. Observation by the author with a 150mm refractor.

made up of around a dozen fairly bright stars and many fainter ones, and makes a lovely low-magnification telescopic sight.

The **Blinking Planetary (NGC 6826)** lies east of **Theta Cyg**. It appears small and well defined, with a slightly blue tinge, its central star visible through a 150mm telescope. Its outer shell appears to blink off only as the observer looks directly at the object, while its bright central star remains visible.

Spread across a portion of the Milky Way in southern Cygnus, the **Veil Nebula** is a supernova remnant, the brightest parts of which, **NGC 6992**, may be discerned through big binoculars from a dark site. Big binoculars will also reveal the **North America Nebula (NGC 7000)**, appearing as a wedge-shaped brightening of the Milky Way (considerably larger than the apparent diameter of the full Moon) to the east of Deneb. Having such a large area and a low surface brightness, it is elusive at higher magnifications through a telescope.

The Veil Nebula in Cygnus, imaged with an astronomical CCD camera.

Northern autumn (fall) stars (midnight, 1 October) ....

After the short, light nights of summertime, autumn (fall) brings with it a dramatic change of celestial scenery. Ursa Major now paces the northern horizon, the Plough swinging low, while the W of Cassiopeia climbs high overhead, cut through by the great frieze of the Milky Way, which arches across the sky from east to west.

Hercules begins to step down towards the northwestern horizon after sunset, followed by Vega, which leads the Summer Triangle's descent. The Northern Cross is standing upright, or, if the traditional constellation image is pictured, the swan plummets head downwards towards the horizon.

While one hero departs in the west, another commences its ascent in the east after sunset. Orion has just about cleared the horizon, his brilliant and easily recognizable hunter's form pursuing red-eyed Taurus. Walking hand-in-hand with each other, the twins Castor and Pollux in Gemini stroll into the autumn (fall) skies; while Auriga's chariot thunders above them, its alpha star Capella shining as a prominent beacon high above the eastern horizon.

Looking south, a large expanse of sky is filled with relatively faint stars and ill-defined star patterns. Although these autumn (fall) constellations may lack the immediate visual impact of many other parts of the sky, they are, nonetheless, packed full of interest. Our view into Sculptor, near the southern horizon, is at right angles to the plane of the Milky Way, and we're peering into intergalactic depths unobstructed by most of the dust and gas of our home Galaxy.

Given a clear horizon, the bright Fomalhaut in Piscis Austrinus flashes its presence in the celestial Water that runs across the southern sky from Eridanus in the southeast, through Cetus, Pisces and Aquarius, into Capricornus in the southwest. Climbing in the east, orange Aldebaran is preceded by the Pleiades in Taurus; to its west is the familiar little pattern of Aries, and next to it the ill-defined pattern of stars making up Pisces, traceable around the bottom left corner of the Square of Pegasus. Andromeda and Triangulum soar high, their boundaries containing the two most distant objects visible with the unaided eye – the Great Spiral in Andromeda and the Pinwheel Galaxy.

Horizon from New York (41°N)

Horizon from London (52°N)

Northern autumn (fall) sky, looking due south (east at left, west at right) from the horizon to the zenith. The horizon lines for London (52°N) and New York (41°N) are marked, as well as the ecliptic. The chart is relevant for 1 August (4am), 1 September (2am), 1 October (midnight), 1 November (10pm) and 1 December (8pm).

Ca

Pollux

Betelg

Procyon

Siri

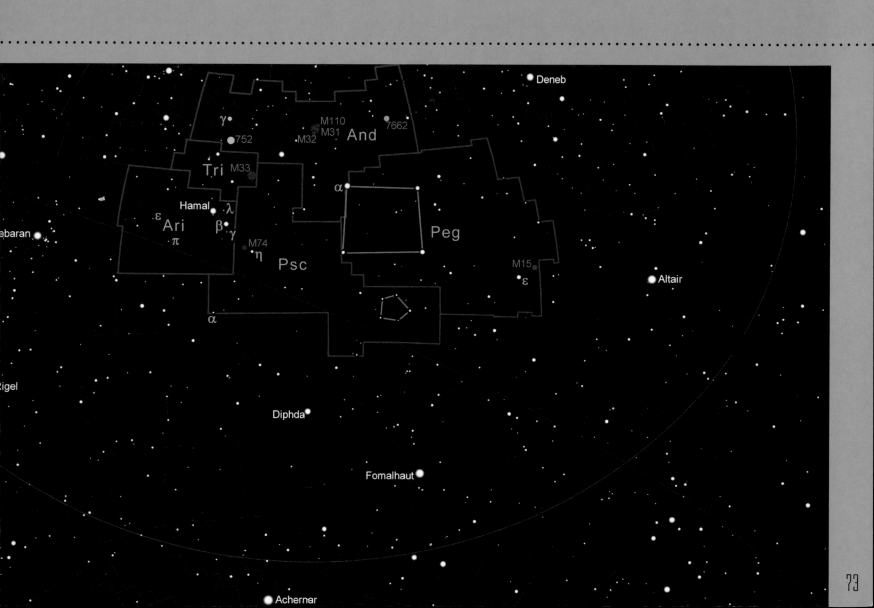

Pegasus ....

PEG / PEGASI
Highest at midnight: early September

Pegasus is a wide constellation, easy to locate because the prominent **Square of Pegasus** asterism takes up much of its eastern parts – but note that the top left star actually belongs to Andromeda. Seeing how many stars are visible within the square is a good test of the darkness of your observing site; if you can spot six stars, then you have a nice dark site.

Epsilon Peg is a lovely wide double star comprising an orange magnitude 2.4 primary and a blue magnitude 8.4 companion. Around four degrees to its northwest lies **M15**, a nice globular cluster bright enough to be seen through binoculars; its outer stars are resolvable through a 150mm telescope.

Globular cluster M15 in Pegasus, observed by the author using a 125mm refractor.

Andromeda ....

AND / ANDROMEDAE
Highest at midnight: late September

Andromeda covers a width of sky about equal to two outstretched hands placed thumb to thumb. Although it's not a particularly brilliant constellation, Andromeda is easily found, since its main star, **Alpha And (Alpheratz)** is the top left-hand star of the Square of Pegasus asterism. Eastward from Alpheratz, Andromeda's brighter stars trace out a widening cutlass-blade shape.

Gamma And (Almach) is a lovely double star with gold and blue components of magnitudes 2.3 and 4.8, easily separable through a small telescope. The fainter component also has a close blue magnitude 6.6 companion which will become resolvable through a 200mm telescope around 2020.

Andromeda is best known for being home to the **Great Andromeda Galaxy (M31)**, the largest member of the Local Group of galaxies, more than 2.5 million light years away. It is easy to locate and can be seen without difficulty with the unaided eye from dark suburban sites. We view M31 from an angle of around 30 degrees above its plane, so that it is somewhat foreshortened. Through binoculars it appears as a bright misty oval around half a degree wide, and from dark sites it stretches yet further across the field of view. A 200mm telescope will reveal hints of structure within the galaxy, including a prominent dark lane and a suggestion of knottiness (large nebulae) in its spiral arms. Nearby are its small satellite galaxies, **M32** and **M110**, both visible as tiny condensed blobs through 80mm binoculars. M32, the brighter of the pair, lies around half a degree south of M31's centre, while M110 lies around a degree to the northwest.

NGC 752 is a large open cluster made up of more than 60 faint stars spread pretty evenly over an area larger than the full Moon. The cluster is visible as a misty patch through binoculars, and its individual stars are resolvable through a 100mm telescope.

The **Blue Snowball (NGC 7662)** is a bright, ninth-magnitude planetary nebula. It is visible through small telescopes as a fuzzy blue spot, and delightful to view through larger instruments, though its colour may not be so apparent at high magnifications.

Comparison between an astronomical CCD image and a visual observation of the Great Spiral in Andromeda (plus its companion galaxies M32 and M110). Field of view is two degrees (four times the apparent width of the Moon).

Pisces

♓ PSC / PISCIUM
Highest at midnight: early October

Lying to the immediate south and east of the Square of Pegasus, Pisces is one of the largest of the 12 Zodiacal constellations. Its traditional outline, comprising a series of faint stars, is only traceable from dark sites. Pisces' brightest star, **Alpha Psc**, lies in the far southeastern corner of the constellation. Through a 100mm telescope it can be resolved as a close double of magnitudes 4.2 and 5.2. In the western corner of Pisces, the well-known asterism of the **Circlet** is made up of seven stars – a challenge to spot with the unaided eye from an urban site.

The Phantom (M74), a face-on spiral galaxy, is Pisces' brightest deep-sky object. It can be found a little more than one degree east of **Eta Psc**, and appears as a sizeable round smudge with a bright, well-defined nucleus through a small telescope.

The Phantom (so called because of its low surface brightness), a face-on galaxy M74 in Pisces, imaged using a 127mm refractor and astronomical CCD camera.

▽ TRI / TRIANGULI
Highest at midnight: late October

Wedged between Aries and Andromeda, Triangulum is one of the sky's smallest and least prominent constellations. It is made up of three faint stars that form an elongated triangle. Insignificant though Triangulum appears, it hosts one of the nearest galaxies – the **Pinwheel Galaxy (M33)** – a face-on spiral some 2.7 million light years distant, just visible with the naked eye from a dark site. This galaxy has a low surface brightness, so although it may be seen through binoculars, it may be missed using a telescope with a higher magnification.

The Pinwheel Galaxy in Triangulum, a face-on, low surface object. Imaged with a 127mm refractor and astronomical CCD camera.

♈ ARI / ARIETIS
Highest at midnight: early November

Aries, the smallest constellation of the Zodiac, can be identified by the small pattern of its brighter stars – **Alpha (Hamal) Ari, Beta Ari** and **Gamma Ari** – which lie some distance west of the Pleiades in neighbouring Taurus. The Sun, Moon and planets are frequent visitors to the southern part of Aries, since this is where a short section of the ecliptic lies.

Gamma Ari is one of the best identical stellar duos in the sky; a double of white magnitude 4.6 stars, easily visible through small telescopes and looking like a pair of glowing eyes. **Lambda Ari** is another wide double, with a white magnitude 4.8 primary and a yellow magnitude 7.3 companion. More challenging doubles in Aries include **Epsilon Ari**, a close pair of white stars of magnitudes 4.6 and 5.5, just separable with a 100mm telescope; and **Pi Ari**, a blue magnitude 5.2 star with a close yellow magnitude 8.5 companion, separable with a 60mm telescope. The constellation harbours no conspicuous deep-sky objects, but ardent deep-sky hunters may attempt observing a smattering of twelfth-magnitude galaxies that lie within Aries' borders.

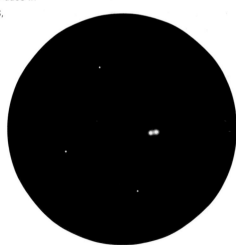

Gamma Ari, a twin double star, observed through a 100mm refractor.

In this section we take a look at the constellations, stars and celestial showpieces of the southern skies, beginning with constellations around the south celestial pole and then by taking a season-by-season view. Far southern stars had to wait until the early 17th century to be mapped by European explorers.

Southern circumpolar constellations

Sigma Octantis, just visible with the naked-eye, marks the approximate location of the south celestial pole. Fortunately there are several convenient pointers to the pole. The nearest consists of the squat triangle formed by Nu Octantis, Beta Hydri and Beta Octantis; the latter is the triangle's apex and points towards the pole about a hand's width away. Adjacent to the triangle are the Small and Large Magellanic Clouds, which themselves form a triangle with the south celestial pole.

On the other side of the pole, the stars Gacrux and Acrux in the Southern Cross also point polewards. Crux itself has its own signpost – the nearby bright circumpolar stars Alpha Centauri and Beta Centauri. Crux is sometimes confused with the False Cross, an asterism comprising Delta Velorum, Kappa Velorum, Iota Carinae and Epsilon Carinae and whose stars don't point to the pole.

Southern circumpolar skies are divided into two distinct halves. One half contains the star-spangled Milky Way running through the constellations of Carina, Musca, Crux, Centaurus and Triangulum Australe. Along this band can be found a wonderful assortment of star clusters and nebulae, including the big, bright Carina Nebula, the giant globular cluster Omega Centauri and the glorious Jewel Box in Crux.

The other half of the circumpolar sky is less overtly spectacular and dotted with a handful of bright stars, notably Canopus in Carina and Achernar in Eridanus. Huddled around Octans at the pole are the constellations of Pavo, Apus, Chamaeleon, Mena, Hydrus, Tucana and Indus, all of which are inconspicuous but traceable from a dark site.

More than compensating for the dearth of naked-eye stellar spectacle is the presence of the Small and Large Magellanic Clouds, delightful irregular galaxies that soar high above the southern horizon during the autumn (fall). Each of these near-galactic neighbours is magnificent through binoculars and telescopes, and they contain many deep-sky treasures, including the spectacular Tarantula Nebula.

Circumpolar limit from Canberra (35°S)

Circumpolar limit from Wellington (41°S)

A broad view of the southern circumpolar sky, looking due south (east at left, west at right). The outer circle represents extent of circumpolarity from Wellington (41°S) and the inner circle for stars that are circumpolar from Canberra (35°S). The ecliptic is also shown at either side (none of it is circumpolar). The chart is relevant for 1 November (4am), 1 December (2am), 1 January (midnight), 1 February (10pm) and 1 March (8pm).

● Arcturus

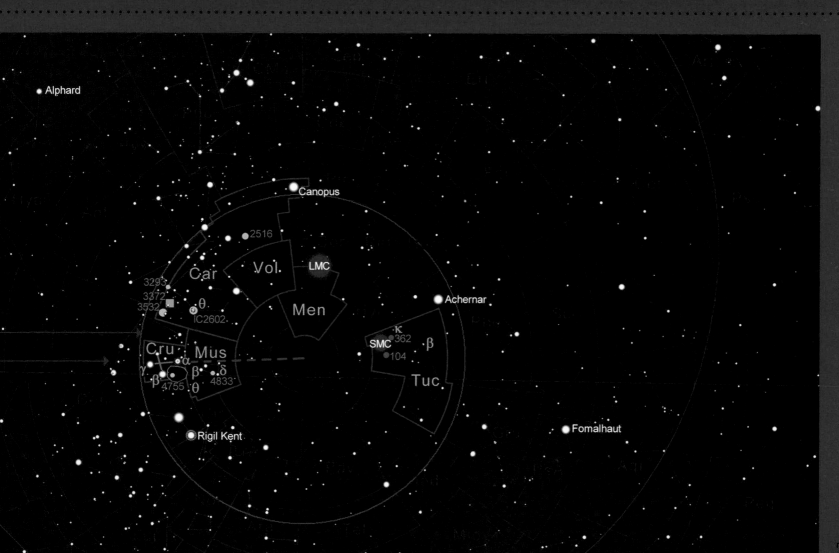

Alphard

Canopus

2516

Vol.

LMC

Car

Men

Achernar

3293
3372
3532

θ

IC2602

κ

362

SMC

β

104

Cru

Mus

Tuc

α

γ

β

β

δ

β

4755

θ

4833

Rigil Kent

Fomalhaut

TUC / TUCANAE
Highest at midnight: mid-September

Tucana is one of the most interesting southern circumpolar constellations. Its main stars are not bright and it is a challenge to trace with the unaided eye, but its location can be identified easily, since it encompasses the **Small Magellanic Cloud (SMC)** and a swathe of sky to its north and northwest. Binoculars will show that **Beta Tuc** comprises beautiful twin blue stars of the fourth magnitude. Another noteworthy double, **Kappa Tuc**, consists of a magnitude 5.1 primary and a magnitude 7.3 companion, divisible through a small telescope.

The SMC, a small irregular galaxy just 200,000 light years from our own Milky Way, appears to the naked eye as a glowing oval patch about two fingers in width. It is a splendid object to scan through binoculars or small telescopes. On its southern edge lie two bright globular clusters. **47 Tuc (NGC 104)** is a very large globular cluster, visible without optical aid and glorious through binoculars. About the same apparent diameter as the full Moon, it is almost as grand as Omega Centauri. Nearby lies **NGC 362**, another beautiful, bright globular cluster visible with the naked eye. Both globulars lie much closer to the Milky Way than to the SMC, so they are very much foreground objects.

Wide-angle view of the southern Milky Way (at right), showing the Magellanic Clouds (at left). Imaged from New Zealand with a digital SLR, combination of five five-minute undriven exposures.

MEN / MENSAE
Highest at midnight: mid-December

Mensa contains fewer bright stars than any other constellation – all of its stars are of the fifth magnitude and below, so it is a real challenge to spot any of them with the unaided eye.

It is easy to pinpoint the space it occupies, as the constellation stretches from the southern portion of the **Large Magellanic Cloud (LMC)** to a few degrees short of the south celestial pole. The portion of the LMC that it covers is scattered with numerous small, faint open clusters, most enjoyable to scan through a telescope at medium magnifications.

The Large Magellanic Cloud, photographed by astronaut Donald Pettit during his spaceflight onboard the International Space Station (ISS).

CRU / CRUCIS
Highest at midnight: late March

Although Crux, or the Southern Cross, is the smallest constellation, it packs a lot of punch. Its four brightest stars – **Alpha Cru, Beta Cru, Gamma Cru** and **Delta Cru** – make a prominent cruciform pattern. Separated by six degrees, Alpha Cru and Gamma Cru point towards the south celestial pole, which lies 27 degrees away in Octans. The Milky Way forms a glamorous backdrop for the entire constellation.

A small telescope can resolve blue Alpha Cru into fairly close twin components of magnitudes 1.3 and 1.8. Another noteworthy double, **Mu Cru**, consists of a wide pair of magnitudes 4 and 5.1.

The **Coalsack Nebula** is a well-defined patch of darkness. some four degrees wide, that occupies the entire southeastern quadrant of Crux, intruding slightly into Musca and Centaurus. Though it gives the impression of being a hole through the Milky Way, it is actually an isolated cloud of dust silhouetted against the bright Milky Way background. Visible as a misty spot with the unaided eye, just north of the Coalsack, shines the **Jewel Box (NGC 4755)**, a marvellous star cluster around **Kappa Cru**. Small telescopes will show a dazzling selection of fairly bright stars of a variety of colours, the brightest of which form a pattern like a miniature Orion's Belt.

The Southern Cross, with the dark Coalsack Nebula, photographed with a driven digital SLR.

MUS / MUSCAE
Highest at midnight: late March

Looking more like a smaller version of Grus than the insect it portrays, Musca is an easily identifiable collection of fairly bright stars immediately south of Crux. Spanning a bright section of the Milky Way, Musca is delightful to peruse through binoculars. **Beta Mus** is a close blue pair of magnitudes 3 and 4, separable using a 100mm telescope. A somewhat easier double, **Theta Mus**, comprises a blue pair of fifth and seventh magnitudes.

Around half a degree north of **Delta Mus** lies the **Southern Butterfly (NGC 4833)**, visible through binoculars as a small fuzzy spot and comfortably resolvable using a 150mm telescope. Several faint loops of unresolved stars emanate from either side of the cluster, giving it the appearance of a butterfly. Compare it with the Butterfly Cluster (M6) in Scorpius – the two are often visible high in the sky at the same time.

Sparkling stellar jewels in the Southern Butterfly in Musca, imaged up-close by the Hubble Space Telescope.

Carina

CAR / CARINAE
Highest at midnight: early February

Carina is a large, wide constellation spanning around 40 degrees of the southern circumpolar regions from its brightest star **Alpha Car (Canopus)** in the west, to its starry eastern realms that bathe in a glorious section of the Milky Way. The outline made by Carina's brighter stars meanders south of Vela; here can be found the **False Cross**, an asterism comprising stars of Carina and Vela, which is sometimes mistaken for the constellation of Crux.

Eta Car is an unpredictable variable which has flared to magnitude -1 in the past. Eta Car is immersed within the **Keyhole Nebula (NGC 3372)**, a very large diffuse nebula easily visible with the unaided eye. With a width of some two degrees – four times the apparent diameter of the Moon – NGC 3372 is one of the most magnificent deep-sky objects in the southern sky. Through binoculars and small telescopes it displays considerable structure, with delicate wisps streaming from discrete clumps of bright nebulosity, interwoven with numerous dark lanes and peppered with dozens of stars. The keyhole refers to the shape of a dark zone within the nebula near Eta Car, not the nebula's general shape.

Several other deep-sky objects can be glimpsed with the unaided eye in the immediate vicinity of Eta Car. Open cluster **NGC 3532** lies just two degrees to its east – a beautiful big agglomeration of stars in the eighth- to twelfth-magnitude range, one of the finest open clusters in the heavens. **X Car**, a third-magnitude yellow supergiant that actually lies far in the cosmic distance, can be seen on the cluster's eastern edge. Nearby, **NGC 3293** is a rather smaller open cluster, remarkable for its mixture of bright blue and red stars. Some distance to the south, the **Southern Pleiades (IC 2602)** dazzles stargazers with its collection of bright stars spread around **Theta Car**, a number of which are visible with the unaided eye.

NGC 2516, in the western part of Carina, is another noteworthy open star cluster visible without optical aid, covering an area the size of the full Moon. Binoculars show that the brightest of its 80 or so stars are arranged in a striking cross shape, appearing particularly concentrated at the intersection of both axes.

The glorious nebula surrounding Eta Carina, including NGC 3372 and NGC 3532.

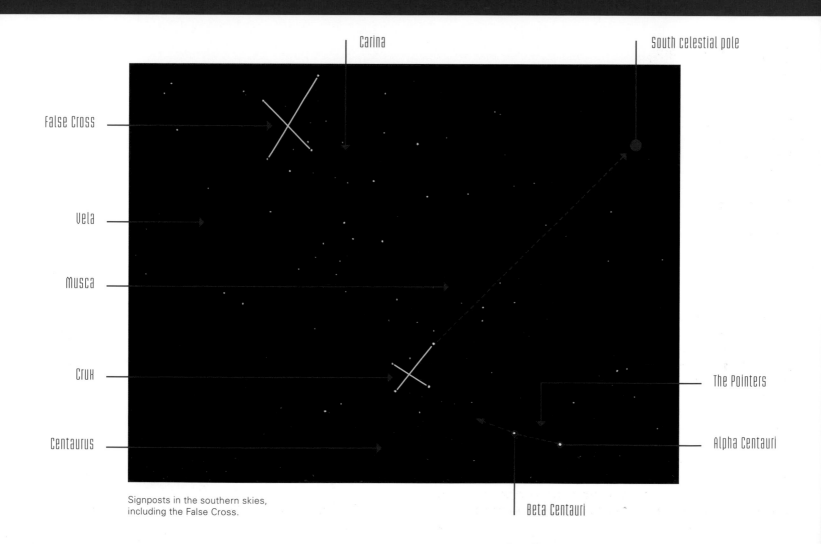

Carina

South celestial pole

False Cross

Vela

Musca

Crux

Centaurus

The Pointers

Alpha Centauri

Beta Centauri

Signposts in the southern skies,
including the False Cross.

Southern summer stars (midnight, 1 January) ....

Summertime skies over the southern hemisphere see the Milky Way climb at a steep angle to the northern and southern horizon. Packed with star clusters and nebulae, the Milky Way stretches across the skies from Auriga, now sinking in the northwest, through Puppis at its most neck-craning heights near the zenith and down into Norma, rising above the southeastern horizon.

Brilliant Canopus in Carina is at its highest in the south, just a hand's width from the zenith. High in the southern sky, between Canopus and the pole, the Large Magellanic Cloud – the Milky Way's biggest satellite galaxy – also enjoys its highest point above the horizon. Meanwhile its less spectacular galactic sibling, the Small Magellanic Cloud, is to the right of the south celestial pole and slowly falling, along with nearby Achernar in Eridanus.

Pavo is low in the south, Alpha Pavonis only a few degrees above the horizon, while below it Telescopium barely manages to show its southern half. Indus, Grus, Tucana and Phoenix follow hard on their heels in their descent towards the circumpolar doldrums. Fomalhaut in Piscis Austrinus drops in the southwest, followed by Sculptor and the large expanse of Cetus in the west. At the same time, the bright stars of Alpha Centauri, Beta Centauri and Crux are climbing ever higher in the southeast and almost the whole of Centaurus has cleared the horizon. Above it, glorious Vela and Carina are on the ascendancy, high in the southeast.

Northern skies are at their most star-studded, with Orion riding high, followed by Canis Major with brilliant Sirius, the sky's brightest star, and Procyon and Canis Minor. Aries, Taurus and the Pleiades are low in the northwest, followed by Castor and Pollux in Gemini and then by Cancer and Leo, rising in the northeast. It is just possible to see Capella in Auriga skim the northern horizon. Columba, preceded by faint Pictor and Caelum, takes zenithal position. Southwest of Orion, the relatively unstarry Eridanus and Fornax take up great swathes of the skies, followed by giant Cetus which is beginning to plunge into the western horizon. In the east, Hydra snakes across the sky, just about to clear the horizon in its entirety.

Horizon from Canberra (35°S)

Horizon from Wellington (41°S)

Southern summer sky, looking due north (west at left, east at right) from the horizon to the zenith. The horizon lines for Wellington (41°S) and Canberra (35°S) are marked, as well as the ecliptic. The chart is relevant for 1 November (4am), 1 December (2am), 1 January (midnight), 1 February (10pm) and 1 March (8pm).

Eridanus

≀≀≀ ERI / ERIDANI
Highest at midnight: late November

Stretching from the celestial equator to a declination of -58 degrees, Eridanus is a large constellation whose stars meander from Taurus's southern border to the northern border of Hydrus. Magnitude 0.5 **Alpha Eri (Achernar)**, a hot blue star, is far and away the brightest in the constellation. Binoculars reveal a line of nine stars – all designated Tau, but spread across 18 degrees. It's only possible to view the entire constellation from locations south of 32°N.

Theta Eri, a lovely blue double of magnitudes 3.2 and 4.3, can be split with a small telescope. Omicron 2 Eri comprises a yellow magnitude 4.4 primary partnered by a magnitude 9.5 white dwarf, the latter having an eleventh-magnitude red dwarf companion; it can be enjoyed through a 100mm telescope. Eridanus's nicest coloured double is **32 Eri**, a wide pair consisting of a yellow magnitude 4.8 primary and a turquoise magnitude 6.1 companion.

Despite its size, Eridanus has few bright deep-sky objects. One of the best, **Cleopatra's Eye (NGC 1535)**, is a lovely bright planetary nebula four degrees east of **Gamma Eri**. A small telescope shows it as a ninth-magnitude greenish ball and some structure, including its central star, can be seen through a 150mm telescope.

Cleopatra's Eye, a bright planetary nebula in Eridanus.

Lepus

LEP / LEPORIS
Highest at midnight: mid-December

Located at the foot of Orion, Lepus is a fairly bright constellation; its broad bow-tie pattern climbs high from southern locations, but it is prominent enough to be easily traced from northern temperate zones too. A small telescope will reveal **Gamma Lep** to be a double star, comprising a yellow magnitude 3.6 primary and an orange magnitude 6.2 companion.

NGC 2017, a lovely group of stars, lies just one and a half degrees east of **Alpha Lep**. A small telescope will show six stars, four of which stand out. Closer scrutiny with a large telescope at a high magnification will reveal that a couple of these stars are close doubles. In the south of Lepus, set in a rich starfield, the compact globular cluster **M79** is a superb sight; many of its brighter stars are resolvable through a 200mm telescope, even near its core, which isn't as bright or as condensed as those of most globulars.

Globular cluster
M79 in Lepus.

Columba

COL / COLUMBAE
Highest at midnight: late December

Although Columba is a small, inconspicuous constellation, it is located between Sirius and Canopus and hence relatively easy to find. Its main stars follow a sinuous east-west path; the brightest of them, third magnitude **Alpha Col (Phakt)**, is a B-type supergiant with a faint binary companion. **Mu Col**, an O-type star just visible with the unaided eye, is quite remarkable; known as the 'runaway' star, it is thought to have been launched out of the Trapezium Cluster in the Orion Nebula by a gravitational 'slingshot' more than two million years ago (along with AE Aur and 53 Ari).

Columba is a deep-sky connoisseur's constellation that requires dark skies and a large telescope to appreciate fully, but a number of objects are accessible using small telescopes. NGC 1851, a compact seventh-magnitude globular cluster resolvable to the core through a 150mm telescope under ideal conditions, is to be found in the constellation's southwestern corner. One of the sky's prettiest little open clusters, **NGC 1963**, can be found less than three degrees southwest of Phakt. Composed of around a dozen stars between the eighth and eleventh magnitude, this loose open cluster is shaped like the number three (or a bow, according to John Herschel); around ten arcminutes across, NGC 1963 lies close to the thirteenth-magnitude edge-on galaxy **IC 2135**. Faint galaxies abound in Columba, the tenth-magnitude duo **NGC 1792** and **NGC 1808** near the constellation's western edge being the brightest among them.

Columba is well known among astronomers for being home to the solar antapex – the point in the sky that the Sun, attended by its family of planets, asteroids and comets, is heading directly away from at a velocity of around 60,000km per hour (37,300mph). The solar apex – the point towards which we're appearing to move – lies on the opposite side of the celestial sphere, in Hercules, just southwest of Vega.

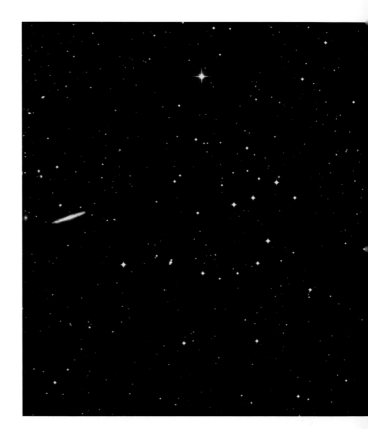

Monoceros ....

MON / MONOCEROTIS
Highest at midnight: early January

A sprawling constellation east of Orion, set against a portion of the Milky Way, Monoceros may be untraceable from urban locations, as its main stars are faint. **Beta Mon** is a dazzling triple, comprising blue stars of magnitudes 3.8, 5 and 5.3, and is easily resolvable through a small telescope. **8 Mon** is a line-of-sight double comprising yellow magnitude 4.4 and blue magnitude 6.7 stars.

S Mon, a bright blue star with a magnitude 7.6 companion, lies within the **Christmas Tree Cluster (NGC 2264)**, a bright cluster easily visible through binoculars. To the southwest lies the **Rosette Nebula**, the faint backdrop for the bright star cluster **NGC 2244**, numbering a dozen or more stars. It is possible to glimpse the Rosette through big binoculars, but it may not be obvious at all through a much larger telescope at higher magnifications.

    The **Heart-Shaped Cluster (M50)** is a sizeable grouping of around 30 fairly bright stars, with an equal number of fainter ones, easily resolvable through a 150mm telescope. **NGC 2353**, a smaller cluster, can be found in the same binocular field, around three degrees to the southeast.

The Rosette Nebula in Monoceros, imaged with a 100mm refractor and astronomical CCD camera (filters used).

Canis Major

CMA / CANIS MAJORIS
Highest at midnight: early January

Sirius, the sky's brightest star. The spikes are due to optical effects caused by the telescope.

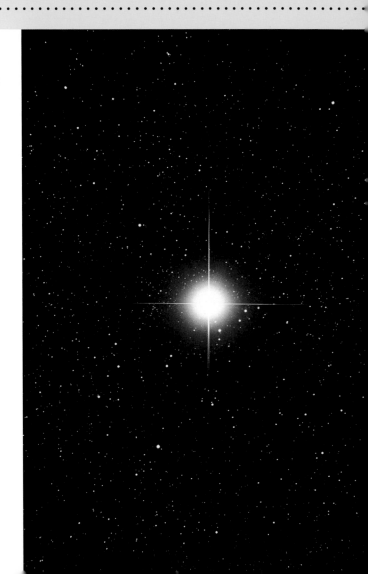

Canis Major is one of the brightest and most easily recognizable constellations. **Alpha CMa (Sirius)** shines at a dazzling magnitude -1.5 and is the sky's brightest star; it can be found by tracing the line of Orion's Belt to the southeast. When seen high in the sky during southern summertime, Sirius is a dazzling steely-white object. From northern temperate locations Sirius never rises very high, and its colourful scintillations are often striking; their wildness is an indication of the quality of atmospheric seeing.

Sirius is a famous binary with a faint white-dwarf companion, **Sirius B**. Orbiting each other in a period of 50 years, the pair was at its closest in the 1990s. Their separation is now increasing and they will be at their widest in the first half of the 2020s, around 10 arcseconds apart – easily resolvable through a small telescope. However, the brilliance of magnitude -1.5 Sirius' brilliance makes spotting eighth-magnitude Sirius B a real challenge. One trick is to follow Sirius as it disappears behind a distant solid object such as a wall – once its light has faded, Sirius B ought to be visible for a moment to its immediate east before it too vanishes.

Some four degrees south of Sirius lies the **Little Beehive (M41)**, a bright open cluster just visible with the unaided eye as a fourth-magnitude smudge. A grand object in a low-to-medium power field, its dozens of stars visible through a small telescope take up an area about as wide as the full Moon, and many of them are arranged along winding lines or clumped into pairs.

Canis Major's eastern sector is occupied by the Milky Way, and in its midst can be found the lovely bright open clusters of **NGC 2360** and **NGC 2362**. NGC 2360, made up of dozens of evenly scattered sub-eleventh magnitude stars bisected by a meandering line of its brighter members, is superb viewed at medium magnification through a 150mm telescope. NGC 2362 contains around 60 stars, which are huddled closely around the fourth-magnitude **Tau CMa**.

Puppis ....

PUP / PUPPIS
Highest at midnight: mid-January

Puppis, a large constellation that lies to the east and south of Canis Major, spreads south across 40 degrees of declination to the northern border of Carina. Crossed by the Milky Way, Puppis is packed with bright open clusters, and binocular sweeps of this area can occupy many pleasant hours of stargazing.

L1 Pup and **L2 Pup** are an optical double, easily separated with the naked eye. L2 Pup is a giant red variable ranging from the second to the sixth magnitude in around 141 days.

Just one degree apart in the north of Puppis, **M46** and **M47** are a spectacular pair of open clusters, easily visible in the same binocular field. M46 can be resolved into dozens of stars through a small telescope, and many more can be seen through a 150mm telescope in its central zone. M47 is the brighter of the duo, and can just be glimpsed with the unaided eye. Dozens of bright stars, all around the same brightness, can be seen through a small telescope. Its central star is an easily resolvable double. **M93**, some distance due south of the pair, is easy to find near **Xi Pup**, but it is somewhat smaller and less spectacular. **NGC 2477,** an open cluster west of the startling blue **Zeta Pup**, can just be discerned with the unaided eye. Twice the diameter of the full Moon, it contains many bright colourful stars – a real delight through binoculars or a telescope at low magnification.

Open cluster NGC 2477 in Puppis.

Southern autumn (fall) stars (midnight, 1 April) ....

Rising at a steep angle to the eastern horizon, the Milky Way arches from Scutum and Sagittarius, through Scorpius and Crux high in the south and then falls through Vela and Puppis to Monoceros, low in the west. Musca and Crux have climbed to their highest in the south, followed by Rigil Kentaurus and Hadar, the Pointers in Centaurus.

Meanwhile, Tucana and the Small Magellanic Cloud are low to the southern horizon, and Achernar, near the southern border of Eridanus, trails behind and makes a valiant effort to flash its light through the near-horizon murk. The Large Magellanic Cloud has rotated to the right of the south celestial pole, along with Canopus. Canis Major approaches the western horizon, but Sirius remains a prominent signpost in the evening skies; beneath it, Lepus burrows into the southwestern horizon. In the southeastern skies Sagittarius raises its bow, nudging Scorpius ever higher, accompanied further south by Corona Australis, Telescopium, Ara and Pavo.

At the zenith, sprawling across a broad swathe of sky, Vela and Centaurus oversee the autumn (fall) skies. Looking north, Hydra's extended form spans the celestial vault between Canis Minor and Cancer, both sinking in the northwest, over to Libra, rising high in the northeast. Hydra's brightest star, Alphard, serves as a prominent marker high in the northwest.

The entire northern sky serves as a relatively clear window north of the cluttered starfields of the Milky Way into deep intergalactic space and the galaxy-rich starfields of Leo, Coma Berenices and Virgo, which cover much of the sky's mid-northern aspect.

As Procyon falls towards the western horizon, Regulus has moved past the meridian while Arcturus and Spica are climbing higher in the northeast. Crater, followed by the little keystone asterism of Corvus, are at their highest in the north and it won't be long before Virgo assumes the meridian. Low in the north, Leo Minor can be seen in its entirely, but only the southern parts of Ursa Major and Canes Venatici manage to rise aove the horizon; you will do well to spot Cor Caroli with the unaided eye, just a few degrees high. Brilliant orange Antares is high in the east, and below it most of Ophiuchus has heaved itself up over the horizon.

Horizon from Canberra (35°S) →
Horizon from Wellington (41°S) →

Rigel

Aldebaran

Southern autumn (fall) sky, looking due north (west at left, east at right) from the horizon to the zenith. The horizon lines for Wellington (41°S) and Canberra (35°S) are marked, as well as the ecliptic. The chart is relevant for 1 February (4am), 1 March (2am), 1 April (midnight), 1 May (10pm) and 1 June (8pm).

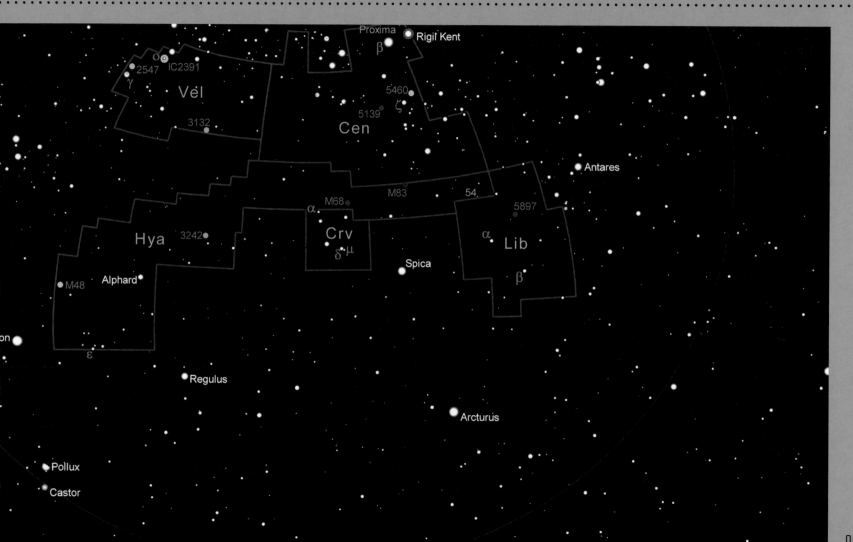

Vela ....

VEL / VELORUM
Highest at midnight: mid-February

Much of this large constellation is immersed in a bright section of the Milky Way north of Carina. Its main outline – a broad oval consisting of a dozen easily visible naked-eye stars – is fairly easy to trace. Once part of a larger constellation called Argo Navis, Vela has inherited its old stellar nomenclature. **Gamma Vel** is its brightest star (there is no Alpha), an interesting multiple star whose main components – blue stars of magnitudes 1.8 and 4.3 – are easily divisible through a small telescope.

Vela billows against a glowing galactic ocean, from which shine numerous open star clusters. It's easy for any stargazer to become utterly absorbed in the beautiful vistas offered by even a small pair of binoculars. **NGC 2547** is a rich treasure trove of around 80 stars covering an area slightly smaller than the full Moon. Various groups and lines of stars within it are easy to discern through a small telescope. Vela's brightest cluster, **IC 2391**, can be seen without optical aid as a misty patch around **Omicron Vel**. It contains about 30 stars visible through a small telescope.

   The **Southern Ring (NGC 3132)** lies on Vela's northern border, a beautiful bright planetary nebula that resembles its northern namesake in Lyra. The nebula's tenth-magnitude central star is easy to spot in a small telescope, and the surrounding field contains several bright stars.

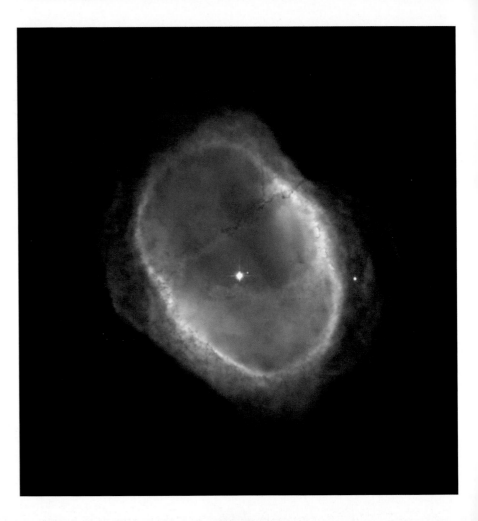

The Southern Ring planetary nebula in Vela,
imaged by the Hubble Space Telescope.

Hydra

HYA / HYDRAE
Highest at midnight: mid-March

Covering more than 1,300 square degrees, Hydra is the largest constellation. Big it may be, but its pattern can be hard to trace. Only the head of Hydra, a prominent clutch of third- and fourth-magnitude stars lying an outstretched hand's width east of the bright star Procyon, rears north of the celestial equator. Southwest from here, the constellation can be traced to its brightest star, magnitude 2 **Alpha Hya**, through various dimmer stars along an undulating path deeper south and east.

Epsilon Hya near the constellation's northwestern border is a binary star with a yellow magnitude 3.4 primary and a blue magnitude 6.7 partner, resolvable with a small telescope. On the opposite side of the constellation, deep in the south, another notable coloured double **54 Hya**, has a yellow magnitude 5.1 primary and a lilac magnitude 7.2 companion.

The **Lawn Sprinkler (M48)** is a large open cluster, easily visible through binoculars. Telescopes will show a spray of fairly bright stars, some of which appear as close pairs. **M68** is a visually interesting bright globular cluster, with intriguing dark lanes and a sizeable dark notch visible through a 200mm telescope. To its east lies the **Southern Pinwheel (M83)**, a face-on barred spiral galaxy, easy to see through binoculars. One of the easiest galaxies in which to discern structure, the spiral shape can be made out through a 150mm telescope. Small telescopes will show the large glowing turquoise disk of the **Ghost of Jupiter (NGC 3242)**, one of the nicest planetary nebulae in the heavens. A 150mm telescope will easily show its eleventh-magnitude central star.

The Southern Pinwheel, a face-on galaxy in Hydra, imaged with a 250mm reflector and astronomical CCD camera.

Corvus

Centaurus

CRV / CORVI
Highest at midnight: late March

CEN / CENTAURI
Highest at midnight: mid-April

Corvus is a small constellation, and although it's not very prominent it can be recognized in a dark sky by its four brightest stars – **Beta Crv**, **Gamma Crv**, **Delta Crv** and **Epsilon Crv**, all of the third magnitude – which form a small trapezium or 'little keystone' asterism to the west of Spica in Virgo.

Alpha Crv is somewhat dimmer and lies two degrees south of Epsilon Crv, while Delta Crv forms an easy naked-eye double with **Eta Crv**, the pair being separated by more than half a degree. Small telescopes will reveal that Delta Crv itself is a double star with a white magnitude 2.9 primary and a rather fainter magnitude 8.5 turquoise companion.

North of Delta Crv, near the northern border of Corvus, lies a small asterism consisting of a large equilateral triangle of stars surrounding a smaller triangle of fainter stars. Known as the **Stargate**, the asterism lies just one degree away from the **Sombrero Galaxy (M104)**, which lies just over the border in Virgo; the two objects can be seen in the same low-magnification, wide-angle field of view.

Planetary nebula **NGC 4361** in central Corvus is a large, almost circular colourless smudge almost one arcminute across; the nebula and its central star are easily seen through a 100mm telescope.

One of the largest constellations, Centaurus's main stars are bright and easily traceable. More stars are visible with the unaided eye in Centaurus than in any other constellation – at least 150 are visible to someone with normal eyesight on a clear dark evening. Binoculars will bring some astounding Centauran starfields into view, particularly along the bright section of the Milky Way in the south.

The remarkable galaxy Centaurus A, imaged with a 250mm reflector and astronomical CCD camera.

Centaurus's brightest stars, the yellow **Alpha Cen (Rigil Kent)** and blue **Beta Cen (Hadar)**, form a lovely pair east of Crux. Just 4.4 light years away, Alpha Cen is the nearest bright star to the Sun. A small telescope will reveal that it is actually a double star, comprising twin components of magnitudes 0 and 1.4. The system also includes a dim (eleventh-magnitude) red dwarf called **Proxima Cen**, located two degrees southwest of Alpha Cen and visible only in the same field of view through a large telescope using a low-magnification, wide-field eyepiece. Proxima takes around one million years to orbit Alpha, and for the next 15,000 years it will be marginally nearer to us, making it the Sun's closest stellar neighbour.

Beta Cen has components of magnitudes 0.6 and 3.9, separable in a 150mm telescope but difficult to identify because of the primary's brightness. Another double, the twin yellow magnitude 2.2 components of **Gamma Cen**, are too close to resolve until after 2020.

To the west of second-magnitude **Zeta Cen**, a large fuzzy third-magnitude star can be discerned with the unaided eye. This is **Omega Centauri (NGC 5139)**, revealed by a small telescope to be the biggest and brightest globular cluster visible in the heavens – a slightly egg-shaped sphere of more than a million stars occupying an area equal to the full Moon. It is majestic through any instrument, but fantastic detail within the dense stellar mass can be discerned through any telescope larger than 150mm. A few degrees north of here can be found NGC 5128, a peculiar giant elliptical galaxy known to radio astronomers as **Centaurus A** because of its strong emissions. Centaurus A appears to have a dark band of dust bisecting it; this actually encircles the galaxy and is seen in silhouette. East of Zeta Cen lies **NGC 5460**, a lovely open cluster on the verge of naked-eye visibility, although none of its 40 or so stars is brighter than eighth magnitude.

LIB / LIBRAE
Highest at midnight: mid-May

One of the smallest and least conspicuous of the Zodiacal constellations, Libra's main stars form a quadrilateral that straddles the ecliptic. It is located northwest of Scorpius, but it is difficult to see with the unaided eye from a city.

Alpha Lib (Zubenelgenubi), a wide double separable in binoculars, consists of a sky-blue magnitude 2.7 star with a white magnitude 5.2 companion. **Beta Lib** is the brightest of Libra's stars and displays an uncommon green hue, the colour being particularly notable through binoculars.

NGC 5897, a small ninth-magnitude globular cluster, is Libra's only noteworthy deep-sky gem. Lying a couple of degrees southeast of **Iota Lib**, it does not have a particularly concentrated core, and its brighter stars can be seen against all of its misty unresolved background – a lovely object viewed through a 200mm telescope at high magnifications.

The awesome globular cluster Omega Centauri, imaged with a 66mm refractor and astronomical CCD camera.

Southern winter stars (midnight, 1 July)......................................

Looking south, Telescopium, Ara, Pavo and Apus soar high above the south celestial pole. The majestic band of the Milky Way forms a steep angle with the horizon as it slices across the sky from Vela in the northeast to southwest, through the centre of our Galaxy in Sagittarius high overhead, to Cygnus in the northeast.

The Large Magellanic Cloud has reached its lowest altitude, and below it, the bright star Canopus only just manages to clear the southern horizon. In the ascendancy are the Small Magellanic Cloud, and Achernar leads the first gushes of Eridanus near its celestial source. Fomalhaut in Piscis Austrinus and Grus, followed by Sculptor and Phoenix, are climbing ever higher in the east. Edging down towards the southwestern horizon, Carina and Vela are followed by Crux, now on its side, and mighty Centaurus with its Pointers, Hadar and Rigil Kentaurus.

Looking north, the ecliptic makes a high arch from Pisces, rising in the east, through Aquarius, Capricornus and Sagittarius near the zenith, via Ophiuchus, Scorpius and Libra to Virgo, slowly sinking in the west. Arcturus in Boötes valiantly battles its decreasing altitude in the northwest. Hercules is at its highest, its northern reaches scraping the horizon, while following it brilliant Vega in Lyra can be seen twinkling a few degrees high. Deneb, one of its northern hemisphere Summer Triangle partners barely clears the horizon and you'd do well to spot it, but Altair,

higher in the sky, fares better. Ophiuchus, Serpens Cauda, Scutum and Aquila cover sections of the Milky Way leading up to the magnificent star clouds, clusters and nebulae of Sagittarius and Scorpius. Crowning the scene, at the zenith, lies Corona Australis. Northeastern skies see Cygnus forlornly struggle to make its presence known as Pegasus nudges tentatively above the horizon.

Horizon from Canberra (35°S) ———⟶

Horizon from Wellington (41°S) ———⟶

Regulus

Southern winter sky, looking due north (west at left, east at right) from the horizon to the zenith. The horizon lines for Wellington (41°S) and Canberra (35°S) are marked, as well as the ecliptic. The chart is relevant for 1 May (4am), 1 June (2am), 1 July (midnight), 1 August (10pm) and 1 September (8pm).

Lupus

LUP / LUPI
Highest at midnight: mid-May

Lying diagonally along the northern edge of the Milky Way, from the east of Centaurus across to the western border of Scorpius, Lupus contains a number of interesting double and multiple stars. **Gamma Lup** is a very close binary with magnitude 3 and 4.4 components that orbit each other every 190 years. At their widest in 1980, they will remain resolvable through a 200mm telescope until 2040. **Eta Lup** is a nice double with a blue magnitude 3.4 primary and a yellow magnitude 7.8 companion, easily resolvable through a small telescope. **Kappa Lup** is a wide double of magnitudes 3.9 and 5.7. The multiple system of **Mu Lup** consists of an easily separable blue magnitude 4.3 primary and a magnitude 6.9 partner. A 200mm telescope will reveal the primary itself to be a close double of twin magnitude 5 stars. **Xi Lup** is a nice pair of blue fifth-magnitude stars, delightful through a small telescope.

Nestled within the bright Milky Way starfields southwest of **Zeta Lup**, the large open cluster **NGC 5822** appears as a Moon-sized hazy patch through binoculars. Small telescopes will resolve its 150 or so loosely assembled stars, among which can be discerned a long, winding stellar chain. **NGC 5986** is the brightest globular cluster within Lupus, small in diameter but resolvable to its bright core using a 200mm telescope.

Scorpius

SCO / SCORPII
Highest at midnight: early June

The False Comet Cluster in Scorpius, imaged with a 250mm reflector and astronomical CCD camera.

Almost overhead in southern winter skies, Scorpius is a spectacular constellation highlighted by the distinctly orange **Alpha Sco (Antares)**. Although low to the horizon when seen from northern temperate locations, the collection of bright stars in the tail of Scorpius near Antares are a familiar sight to northern temperate stargazers (though somewhat dimmed by atmospheric murk). Through a telescope, the fainter blue companion to Antares (an orange supergiant) may be glimpsed through quite a small instrument, although the brilliance of Antares hinders its visibility. West of Antares, the misty patch of globular cluster **M4** can be seen with the unaided eye. A 100mm telescope will resolve it.

The **Butterfly Cluster (M6)** and **Ptolemy's Cluster (M7)** are two magnificent open clusters visible without optical aid. Set like a large lustrous garnet on the eastern wing of the delightful diamond brooch of M6 is the orange giant **BM Sco**. Just three degrees southeast of M6, the much larger and brighter M7 is truly dazzling. Binoculars will reveal many of the brighter stars in this cluster, immersed in the Milky Way. Through a telescope at low magnifications it is most impressive, its central stars forming a distinct H shape. In southern Scorpius, the **False Comet Cluster (NGC 6231)** can be glimpsed with the naked eye as a banana-shaped patch that spreads north from **Zeta Sco**. Through binoculars and small telescopes it appears spectacular, with a core of half a dozen bright stars surrounded by dozens of fainter stars that mesh with the stars of a rich segment of the Milky Way.

The Milky Way in
Scorpius and Sagittarius,
imaged from La Palma
with a driven digital SLR.

Serpens Cauda

SER / SERPENTIS
Highest at midnight: late June

Serpens is a constellation split into two sections on either side of Ophiuchus. **Serpens Cauda**, the constellation's fainter half, is made up of a line of stars that run along the Milky Way's dark central reservation.

M16, an open star cluster comprising a scattering of around a dozen stars, can be discerned through binoculars. Under dark skies, hints of the **Eagle Nebula** are visible in and around the southeast of the cluster.

The Eagle Nebula near M16, imaged with a 127mm refractor and astronomical CCD camera.

Scutum

SCT / SCUTI
Highest at midnight: early July

Scutum may be tiny, and its stars difficult to discern with the unaided eye, but it is packed with deep-sky wonder.

Set against the gorgeous Milky Way background, the **Wild Duck (M11)** is a glorious broad arc of around 200 stars; visible through binoculars as a hazy patch, the cluster is resolvable through a 100mm telescope at higher magnifications. A few degrees to its south lies a fainter and less populous open cluster, **M26**.

The Wild Duck Cluster (upper left), M26 (lower right) and globular cluster NGC 6712, observed by the author using 15x70 binoculars.

Corona Australis .... Sagittarius ....

 ### CRA / CORONAE AUSTRALIS
Highest at midnight: early July

 ### SAG / SAGITTARII
Highest at midnight: early July

Enclosed within a small rectangle of sky that dips into the Milky Way south of Sagittarius, Corona Australis is a pretty little constellation whose brightest stars in the east form a lovely arc. It is not as large or as prominent as its northern counterpart, Corona Borealis, and is best viewed through binoculars.

Kappa 2 CrA and **Kappa 1 CrA** form a lovely blue double comprising a magnitude 5.9 primary and magnitude 6.6 partner, easily separated with a small telescope. **NGC 6541** in the constellation's southwestern corner is a rare example of a globular set within the Milky Way. Visible through binoculars as a seventh-magnitude smudge with a bright core, set within a rich field, much structure can be viewed through a 150mm telescope.

Sagittarius is justifiably considered by many stargazers – especially those in the southern hemisphere, who see it climb high overhead in winter – to be the grandest constellation of them all. It is crossed by a broad swathe of the Milky Way, and beyond it is the centre of our Galaxy, positioned near the constellation's far western border. Sagittarius is full of deep-sky splendours.

The Lagoon Nebula in Sagittarius, imaged with a 250mm reflector and astronomical CCD camera.

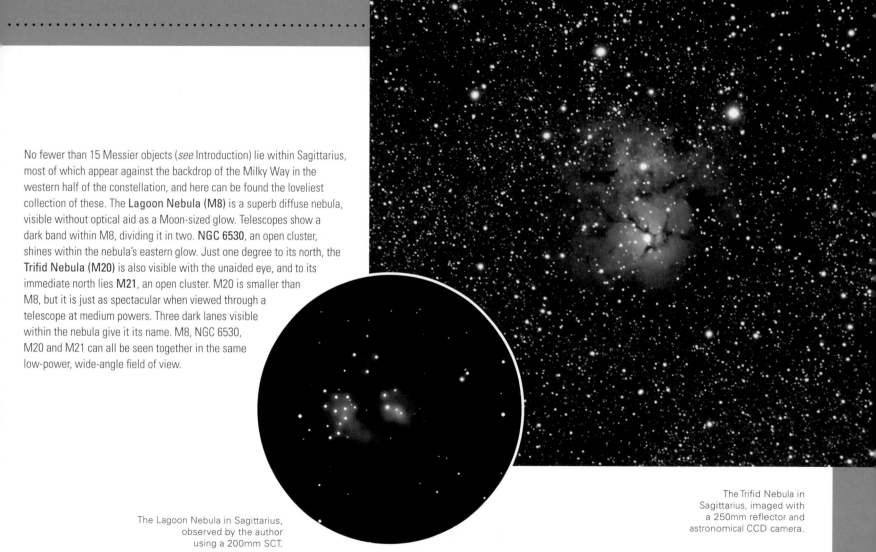

No fewer than 15 Messier objects (*see* Introduction) lie within Sagittarius, most of which appear against the backdrop of the Milky Way in the western half of the constellation, and here can be found the loveliest collection of these. The **Lagoon Nebula (M8)** is a superb diffuse nebula, visible without optical aid as a Moon-sized glow. Telescopes show a dark band within M8, dividing it in two. **NGC 6530**, an open cluster, shines within the nebula's eastern glow. Just one degree to its north, the **Trifid Nebula (M20)** is also visible with the unaided eye, and to its immediate north lies **M21**, an open cluster. M20 is smaller than M8, but it is just as spectacular when viewed through a telescope at medium powers. Three dark lanes visible within the nebula give it its name. M8, NGC 6530, M20 and M21 can all be seen together in the same low-power, wide-angle field of view.

The Lagoon Nebula in Sagittarius, observed by the author using a 200mm SCT.

The Trifid Nebula in Sagittarius, imaged with a 250mm reflector and astronomical CCD camera.

Southern spring stars (midnight, 1 October) ....

Looking south, the Milky Way lies low, almost parallel to the horizon, as the glorious hub of the Galaxy in Sagittarius sinks into the southwest. 'Upside-down', the Southern Cross is at its lowest, a hand's width above the southern horizon, and only the southernmost part of Centaurus remains visible, the Pointers Hadar and Rigil Kentaurus lying to the right of Crux. Procyon rises in the east, and Canopus steadily guides Carina out of the doldrums in the southeastern skies.

At its highest, the Small Magellanic Cloud floats high above the south celestial pole, followed by an ascendant Large Magellanic Cloud. Tucana looms high, followed by Achernar in Eridanus, and the zenithal position is taken by the constellation of Phoenix. Rising in the southeast are the constellations of Vela, Puppis and Canis Major, the ascendancy of brilliant Sirius indicating that southern summer isn't too far off.

Altair in Aquila leads its neighbouring constellations Delphinus and Capricornus as they sink towards the western horizon. The Square of Pegasus is at its highest in the north; beneath it, the Andromeda Galaxy clears the horizon by a few degrees and can be seen through binoculars; its near galactic neighbour, the Triangulum Galaxy, has such a low surface brightness that it is challenging to discern at such a low altitude.

Mid-northern skies are dominated by Capricornus, Aquarius, Cetus and Eridanus, which spread themselves comfortably over much of this aspect of the heavens. Above them, Fomalhaut has risen to prominence. Meanwhile,

the more earthly constellations of Aries, Taurus and Orion now begin to ease themselves above the northeastern horizon. In addition to a fabulous array of galaxies, there are a number of lovely planetary nebulae, globular clusters, double and variable stars to be found amid the spring constellations.

Horizon from Canberra (35°S) →

Horizon from Wellington (41°S) →

A

● Ras

Southern spring sky, looking due north (west at left, east at right) from the horizon to the zenith. The horizon lines for Wellington (41°S) and Canberra (35°S) are marked, as well as the ecliptic. The chart is relevant for 1 August (4am), 1 September (2am), 1 October (midnight), 1 November (10pm) and 1 December (8pm).

Capricornus

♑ CAP / CAPRICORNI
Highest at midnight: early August

A fairly small Zodiacal constellation, and quite an obscure one too, Capricornus comprises a collection of third- and fourth-magnitude stars arranged in a broad south-pointing arrowhead, 'roof' or 'tent' pattern when viewed from the southern hemisphere.

Alpha Cap is a close naked-eye double comprising **Alpha 2 Cap**, a yellow giant of magnitude 3.7 and **Alpha 1 Cap**, an orange supergiant of magnitude 4.3. The pair is an easily separated line-of-sight double, the components lying at 109 and 887 light years away respectively. Each of these stars is itself a wide double with faint companions, the dimmest visible through a 200mm telescope. **Beta Cap** is a nice coloured double with a golden magnitude 3 primary and a sky-blue magnitude 6.1 partner, wide enough to be seen through binoculars.

Capricornus's brightest deep-sky delight is **M30**, a middling sized, fairly bright globular cluster whose outer regions can be resolved through a 200mm telescope. Residing in the constellation's southwestern corner is its brightest galaxy, **NGC 6907**, a beautiful twelfth-magnitude face-on barred spiral, best viewed through telescopes larger than 250mm.

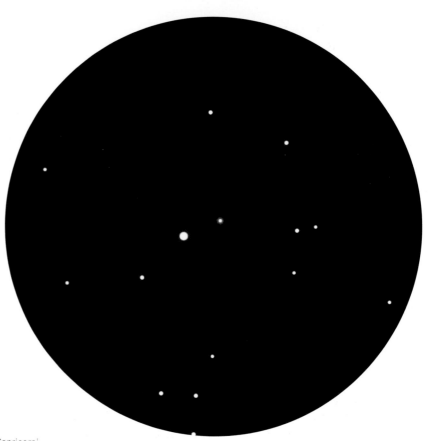

Double star Beta Capricorni,
observed by the author
using a 200mm SCT.

AQR / AQUARII
Highest at midnight: late August

Aquarius is a large constellation whose main pattern comprises a widely spaced collection of second- and third-magnitude stars lying south of the celestial equator. Immediately east of **Alpha Aqr,** a small but striking asterism often called the **Propeller** or the **Water Jar** comprises a pattern of four stars, whose central star, **Zeta Aqr,** is a binary with twin white components of magnitudes 4.3 and 4.4. The system is slowly opening up to us; it can be resolved through a good 80mm telescope, and by the end of the 21st century it will be an easy double through any small telescope.

Three different types of deep-sky object – a planetary nebula, an open cluster and a globular cluster, all visible in the same field of view at low magnifications – can be found in the western reaches of Aquarius. Less than one and a half degrees west of **Nu Aqr** lies the **Saturn Nebula (NGC 7009),** a superb planetary nebula, visible through a 80mm telescope as an elliptical disk of a similar size to Saturn. Larger instruments will reveal the blue colour and some structure, including a narrow inner ellipse and two small lobes protruding from the nebula, giving it the appearance of Saturn with its rings presented edge-on to us. To its southwest is **M73,** a small open cluster shaped rather like the Propeller asterism. Nearby **M72,** a small ninth-magnitude globular cluster with a bright core, can be difficult to resolve using anything smaller than a 250mm telescope.

   M2, the brightest globular cluster in the region, is located less than five degrees north of **Beta Aqr.** It can be seen through binoculars as a fuzzy patch, and a large number of its stars can be seen through a 150mm telescope.

   In the far south of Aquarius, the **Helix Nebula (NGC 7293)** can be seen fairly easily through binoculars as a circular smudge. Although it has the largest apparent diameter of all planetary nebulae, it has a low surface brightness, so the best views are at low magnifications. Hints of its ring structure and some mottling may be glimpsed through a 250mm telescope.

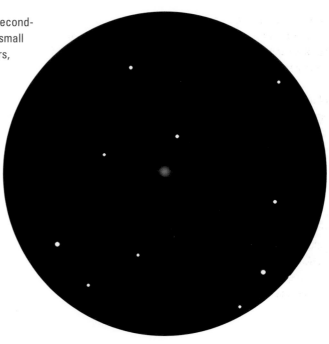

The Saturn Nebula in Aquarius, observed by the author using a 300mm reflector.

The Helix Nebula in
Aquarius, imaged with
a 127mm refractor
and astronomical
CCD camera.

PSA / PISCIS AUSTRINI
Highest at midnight: late August

Piscis Austrinus lies on a cold rectangular celestial fishmonger's slab, and is a rather unspectacular stellar array, save for bright-blue **Alpha PsA (Fomalhaut)**. Shining at first magnitude, Fomalhaut dominates its celestial surroundings, and despite its high southerly declination it is even familiar to those living in northern temperate locations.

Beta PsA is a wide double, resolvable through binoculars, with components of magnitudes 4.5 and 7.5. **Delta PsA** and **Gamma PsA**, immediately south of Fomalhaut, form a wide naked-eye double. Fourth-magnitude Delta PsA has a faint tenth-magnitude companion, visible through an 80mm telescope. Gamma PsA has components of magnitudes 4.5 and 8.5, visible through a small telescope. **Eta PsA** is a very close double of white magnitude 5.4 and 6.6 stars, resolvable through a 100mm telescope.

Fomalhaut in
Piscis Austrinus.

Sculptor........

SCL / SCULPTORIS
Highest at midnight: early October

One of the sky's least conspicuous constellations, Sculptor is most notable for being home to the **South Galactic Pole**. Its whereabouts, east of the bright star Fomalhaut, is easy to locate, but Sculptor's faint stars are challenging to see even when the constellation is high in a dark southern-hemisphere sky.

Positioned very close to the South Galactic Pole, the bright globular **NGC 288** and the splendid nearly edge-on galaxy **NGC 253** are under two degrees apart and can be viewed together in the same low-power field. NGC 253 is bright with an easily discernable nucleus. Through a 100mm telescope considerable structure can be seen along the galaxy's arms, even though it is very foreshortened. NGC 253 lies in a field rich in foreground stars. **NGC 55** is another bright, almost edge-on galaxy, with less obvious visual detail than can be seen in NGC 253.

NGC 253, an edge-on spiral galaxy in Sculptor, imaged with a 127mm refractor and astronomical CCD camera.

CET / CETI
Highest at midnight: mid-October

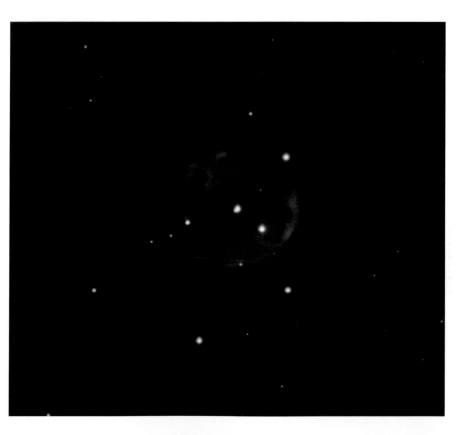

The Skull Nebula in Cetus, imaged with a 127mm refractor and astronomical CCD camera.

A very large constellation lying south of the ecliptic, the bulk of the Whale is immersed south of the celestial equator, but its head rears north and eyes little Aries and glamorous Taurus. Its second brightest star, the red giant **Alpha Cet (Menkar)** makes a wide but charming double with the blue fifth-magnitude star **93 Cet** to its north. **Beta Cet** (magnitude 2), far to the southwest, is the brightest star in Cetus.

Two stars of particular interest to astronomers can be seen with the unaided eye. The red giant **Omicron Cet (Mira)**, is a favourite of variable star observers. It varies between the third and ninth magnitudes over a period of around 332 days. **Tau Cet** is a star remarkably like our own Sun; just 12 light years away, it has an apparent brightness of magnitude 3.5. Observations have revealed that the space surrounding Tau Cet contains ten times more asteroid and comet debris than found in the Sun's vicinity; any planets there are likely to experience high rates of catastrophic collisions.

The **Skull Nebula (NGC 246)**, an eighth-magnitude planetary nebula makes a nice target for close scrutiny with a 200mm telescope. At high magnification it appears as a grey mottled ellipse, with several eleventh-magnitude stars visible in its vicinity, including its central star. **M77**, a tenth-magnitude spiral galaxy with a bright, highly condensed nucleus, can be found by sweeping less than one degree east of **Delta Cet**.

Part Two: The Solar System

A tremendous variety of objects and events within our own Solar System can be viewed by the vigilant stargazer, from amazing phenomena that take place high in our own atmosphere, to the Moon and planets beyond. As the Earth's shadow drapes its curved blanket over the high atmosphere, rich and enchanting colours light up the horizon as longer wavelengths of light are gradually filtered out of the fading sunlight; bright stars and planets begin to emerge from the dwindling afterglow. Noctilucent cloud, aurorae and meteors are sometimes seen as evening progresses, and as night falls we can marvel at celestial objects that are set like jewels against the black velvet of space.

An imaginary but entirely possible evening twilight scene – the Solar System steals the show, with the young crescent Moon, Mercury, Venus, Jupiter and a bright comet in the western skies.

A number of easily visible astronomical phenomena take place within the atmosphere, a layer of gases just 600km (370 miles) deep that separates us from the harsh vacuum of outer space.

Solar spectacles

Sometimes the Sun is encircled by a halo – a luminous ring around 44 degrees in diameter, which often has a reddish inner border and a diffuse outer edge. Occasionally, prominent bright glowing patches can be seen about 22 degrees on either side of the Sun. These are often referred to as sundogs, and they can appear brilliant and colourful. When seen together, a bright solar halo with sundogs on either side is a most impressive sight. Both haloes and sundogs are formed when sunlight is refracted by ice crystals high in the atmosphere (around the same height as commercial jet airliners fly) but each is caused by different shaped ice crystals. The Sun is a million times too bright to look at directly with the unaided eye, so take care when viewing these phenomena by covering the Sun with your hand, or by obscuring it behind a nearby landmark to ensure safe, glare-free viewing. Haloes, caused by the same processes, can also sometimes be seen around the Moon.

A splendid lunar halo.

Noctilucent cloud

Most weather takes place in the troposphere, the lower 15km (9 miles) of the atmosphere. But high above all the regular clouds, at a height of around 85km (53 miles), floats a wispy layer of clouds composed of ice crystals. Being so high, these clouds can remain lit by sunlight long after the Earth directly beneath has been plunged into the shadow of night. This noctilucent cloud ('bright at night') can appear really dramatic when set against a dark evening sky.

Aurorae

A constant stream of energetic particles (electrons and protons) is carried away from the Sun within the solar wind. On encountering the Earth's strong magnetic field, these particles are channelled down into the atmosphere, and when they hit gas atoms within the atmosphere, a luminous glow is produced. This fantastic light show is called the aurora, known as the Aurora Borealis in the northern hemisphere, and Aurora Australis in the southern hemisphere. Colours within aurorae are produced by different glowing gases (oxygen and nitrogen) between 60 and 200km (37–125 miles) high.

Aurorae are produced in a region circling the magnetic poles, but on those occasions when there is a high bombardment of particles from the Sun, they can be seen from locations as far south as the Mediterranean or as far north as Australia, and vivid ones can sometimes be bright enough to be seen from urban locations. Aurorae take on numerous forms, including broad and rayed arcs, flowing curtains and fabulous streaming coronae, all of which appear to change minute by minute as the activity progresses.

Noctilucent cloud observed over Wiltshire, UK, in July 2010 (five image DSLR composite).

The Aurora Borealis, imaged from Iceland.

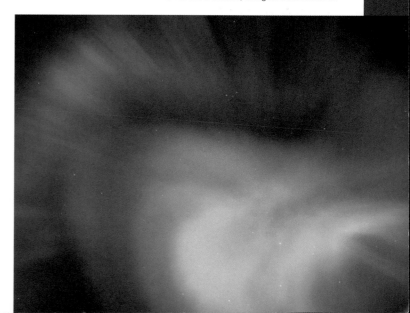

Meteors

A meteor is the glowing trail left when a small meteoroid – a particle of debris left in the wake of a comet – burns up in the Earth's atmosphere at a height between around 75 and 100km (45–65 miles). Though a fleeting phenomenon, the sudden flash and rapid flight of a meteor across the sky is exhilarating to view.

Annual showers

Every year the Earth passes through a number of meteoroid streams, producing annual meteor showers that occur at around the same date each year. Meteors produced by the annual showers appear to radiate from a well-defined point in the sky – an effect of perspective as the Earth passes through the stream – and the radiant is named after the constellation in which it lies. If you watch the area of sky near a radiant during the period of a major shower, you'll be unlucky not to see at least one meteor within a quarter of an hour.

Despite the ominous appearance of the brightest meteors (including fireballs of exceptional brilliance) seen during some of the annual showers, they pose absolutely no threat to the stargazer. Meteoroids within the well-known streams range in size from grapes to grains of sand, and they all burn up completely in the atmosphere. Those objects large and solid enough to survive a superheated descent through the atmosphere originated from asteroids (and a rare few from the Moon and Mars), and they aren't associated with the annual meteor showers.

A brilliant fireball streaks through Leo and past Mars, observed by the author in January 2012.

SOME OF THE YEAR'S BEST METEOR SHOWERS

- **Quadrantids** Active 1–5 January. Maximum on 3 January. The year's first meteor shower emanates from the now-defunct constellation of Quadrans Muralis (the Wall Quadrant), now part of Boötes. Radiant high after midnight. Displays vary, but there are often brief, energetic maxima with high rates. Visible from the northern hemisphere.

- **Centaurids** Active 28 January–21 February. Maximum on 8 February. Comprising two showers – the Alpha Centaurids and Beta Centaurids – with close radiants and peaking on around the same date. Beta Centaurids often outshine the Alpha Centaurids, with some fireballs and persistent trains. Visible from the southern hemisphere.

- **Lyrids** Active 16–25 April. Maximum on 21/22 April. Brilliant medium-speed meteors, some leaving persistent trains. Visible from the northern hemisphere.

- **Pi Puppids** Active 15–28 April. Maximum on 23 April. Variable hourly rate but more active around the time when its parent comet (26P/Grigg-Skjellerup) is nearest the Sun every five years. Visible from the southern hemisphere.

- **Eta Aquarids** Active 21 April–12 May. Maximum on 5/6 May. Very fast meteors with persistent trains. The Eta Aquarids are composed of orbiting debris left in the wake of Halley's Comet. Visible from both hemispheres.

- **Perseids** Active 23 July–22 August. Maximum on 12/13 August. One of the most popular of the annual meteor showers, the Perseids are very fast and often extremely bright, leaving glowing trains. Visible from the northern hemisphere.

- **Orionids** Active 15–29 October. Maximum on 21 October. Bright, very fast meteors, with some fireballs. The stream is formed of debris from Halley's Comet. Visible from both hemispheres.

- **Leonids** Active 13–20 November. Maximum on 17/18 November. Very fast meteors with persistent trains. Enhanced Leonid activity occurs from time to time, with spectacular meteor storms taking place roughly every 33 years as the Earth passes through denser parts of the Leonid stream. Visible from both hemispheres.

- **Phoenicids** Active 28 November–9 December. Maximum on 6 December. Radiant well placed all night. Although activity is unreliable, outbursts have been seen, so they are worth monitoring. Visible from the southern hemisphere.

- **Geminids** Active 6–19 December. Maximum on 13/14 December. Very bright, intensely white meteors, slow moving. Visible from both hemispheres.

Observing meteors

Meteor watching is so enjoyable because it requires no equipment but a keen pair of eyes. A watch of about an hour or two around the dates of the annual shower maxima will satisfy most meteor observers. Observing is best done well away from overt light pollution and on a night when the Moon isn't high and near full phase. It is important to be comfortable. Even during the summer, the nights can be chilly, so wrap up well. Sit on a comfy garden chair, or better still a recliner, facing in the general direction of the meteor radiant. Simple photography using an unguided long time exposure setting will usually pick up one or two brighter meteors during an exposure of ten minutes during periods of good activity.

As the Earth orbits the Sun, the Sun appears to trace a path against the background constellations throughout the year. This path – a projection of Earth's orbital plane on the celestial sphere – is called the ecliptic. All the major planets and most of the minor planets have orbital planes roughly coinciding with that of the Earth, and they therefore all follow paths within a few degrees of the ecliptic plane. The Moon's orbital plane around the Earth lies close to the ecliptic, too.

Being a path common to the Sun, Moon and planets, the ecliptic is a busy thoroughfare. The Moon occasionally passes in front of the Sun, producing solar eclipses. The Moon and planets often appear to approach each other very closely. Sometimes the Moon will move directly in front of a planet, hiding it temporarily and producing an occultation. Close approaches between planets are called appulses, and conjunctions occur when two planets share the same Right Ascension. On very rare occasions one planet will appear to move in front of another, producing a mutual occultation (the temporary hiding of one celestial object behind another more distant object, eg., the occultation of a star or planet by the Moon.); the next one will take place in 2123, when Venus passes in front of Jupiter. The Sun also regularly appears to make close approaches to the planets, but the solar glare renders such events unobservable, apart from those occasions when Mercury and Venus transit the Sun.

The Sun and its family – four terrestrial planets and four gas giants.

Although the Sun is an average-sized star – a 'mere' 1.4 million kilometres (870,000 miles) in diameter – it is far from mundane. The solar surface is a constant turmoil of activity, much of which is easily visible through a small telescope.

Inside the Sun, nuclear fusion converts hydrogen to helium. Each and every second, fusion converts a staggering four million metric tons of gas into energy. A powerful magnetic field is generated within the Sun, and disturbances within it cause sunspots to appear in the photosphere, the Sun's glowing surface. With a temperature of 5,500°C (9,900°F), the photosphere is up to 2,000°C (3,600°F) hotter than a sunspot's interior; such a big contrast in temperature and brightness makes sunspots appear dark against the photosphere.

Solar cycle

Records of solar activity show that the Sun has an 11-year cycle, where the amount of spots seen on the Sun's disc rises and falls. At solar minimum, when the Sun is at its least active, the disc can be spotless for several weeks. Solar maximum sees a proliferation of spots on a daily basis, with occasional giant spots so large that they can be seen with the unaided eye through a proper solar filter. The next solar maximum will take place around 2022.

Observing the Sun

A sunspot first seen at the western edge of the Sun will be carried across to the eastern edge in less than a fortnight. Most sunspots have a lifetime of less than one rotation of the Sun (around 25 days at the Sun's equator), but really big ones – those so huge that they could easily accommodate a dozen Earths – sometimes survive several months. Sunspots have a fascinating structure that morphs from day to day. Their interior (called the umbra) is often dark and featureless, but the area surrounding it (the penumbra) is grey and often striated with a mass of radial lines. Sunspots don't always appear in isolation. Large sunspot groups usually have a main header spot and a slightly smaller follower spot, among a number of others, all embedded within a larger penumbra.

Sunspots are the Sun's most conspicuous feature, but a large telescope under good conditions will reveal a fine structure across the Sun's surface known as granulation, caused by a multitude of bubbling convection cells in the photosphere. Distinctly brighter areas called faculae can sometimes be seen on the photosphere, usually appearing most prominent towards the edge of the Sun where the photosphere is less bright.

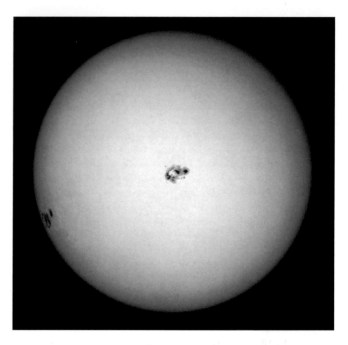

Large groups of sunspots can be seen near the centre of the Sun's disk and about to rotate around the Sun's edge (at left).

SAFETY FIRST

It is dangerous to view the Sun through binoculars or a telescope – a fraction of a second of magnified sunlight can cause permanent blindness. The safest way to observe it is to project the Sun's image onto a shielded smooth white card, but note that some telescopes and eyepieces have plastic parts that may melt if subjected to magnified sunlight. If your telescope has a finder, keep its lenses covered and don't attempt to locate the Sun with it. Don't leave your telescope unattended while observing the Sun, as less experienced people may be tempted to look at the Sun through it. More experienced observers use special solar filters that fit over the telescope aperture, and follow the instructions on their use to the letter. Never use small dark filters that fit on eyepieces, and never use any other household materials as a solar filter, as these are quite inadequate to protect your eyes from the potentially blinding solar radiation.

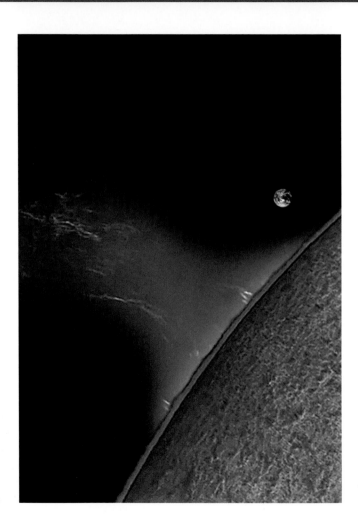

A large ejection of gas from the Sun's surface, known as a prominence, erupts in May 2009. Scale of Earth compared. A 40mm H-alpha solar telescope was used.

SEEING THE SUN IN A DIFFERENT LIGHT

A lot more activity on the Sun can be seen through special filters that block out all light except a band of red light called H-alpha. The chromosphere, a layer of the Sun's atmosphere just above the photosphere, emits H-alpha light, and within it large dense clouds of gas called prominences can be seen, in addition to features visible in ordinary light. Active prominences can appear to change their appearance over just a few minutes, and they can develop into a variety of stunning forms. Like sunspots, prominences are associated with strong magnetic activity. H-alpha telescopes are extremely expensive, and only a minority of stargazers have ever had the pleasure of peering at the Sun through one.

Solar eclipses

It's an amazing coincidence that both the Sun and Moon are around a hundred times their own diameter away from the Earth, and they both appear to have around the same apparent size of half a degree as seen from the Earth. Occasionally the Moon passes in front of the Sun, blocking part or all of the Sun's light from reaching a portion of the Earth's surface.

Partial solar eclipses are seen when the Moon only partially covers the Sun. Even a large partial eclipse isn't safe to look at with the unaided eye or through any non-filtered instrument because even a small chink of direct sunlight can severely damage the unprotected eye. The best way of enjoying partial solar eclipses is to view them through special solar glasses made from aluminized mylar; they are inexpensive and can be bought from any reputable astronomy store. Alternatively, a magnified view of the partially eclipsed Sun can be obtained by projecting the Sun's image through a telescope onto a shadowed white card, held at a suitable distance from the eyepiece (*see* Safety first). Experienced observers also use other methods such as safe whole-aperture filters or various kinds and white light prisms called Herschel wedges.

Sometimes, even when the centre point of the Moon covers the centre point of the Sun, there remains visible a ring of sunlight around the Moon. Known as annular eclipses, they are seen at times when the Moon is around its most distant from the Earth and/or when the Earth is around its closest to the Sun – the darkest part of the Moon's shadow cone fails to make contact with the Earth. Annular eclipses are spectacular in

their own way, but they aren't safe to view without taking the right safety precautions.

The Moon's dark shadow only just touches the Earth under favourable circumstances, when the Earth-Moon-Sun alignment takes place at a time when the Moon's apparent diameter is large enough to hide the entire solar disk, producing a total solar eclipse. Because of the small point of shadow contact with the Earth they can only be seen along a narrow path as the Moon moves through space and the Earth revolves beneath the shadow. On either side of the line of totality lies a broader zone where a partial eclipse is visible, and the further away from the central line the smaller will be the partial eclipse visible.

The total solar eclipse is one of nature's most awesome spectacles. For a few brief moments the Sun is completely hidden as the viewer is plunged into darkness; the atmosphere becomes cooler and a spooky silence pervades the surroundings. Once the Sun is obscured, bright stars and planets become visible in the sky. Here and there the Moon's edge is punctuated by glorious red prominences, and the pearly streamers of the corona, the Sun's outer atmosphere, become visible. It is safe to view the Sun directly only during totality. In rare cases

Totality at the eclipse of 29 March 2006, imaged from Side, Turkey. The delicate pearly streamers of the corona can be seen spreading away from the Sun.

a total eclipse can last more than seven minutes, but most of them have a much shorter duration. As the Moon moves on, sunlight streams through lunar valleys at the Moon's edge and the end of totality is marked by a glorious diamond ring effect.

For billions of years, the Moon has partnered the Earth in its orbit around the Sun. Its silent, doleful face has witnessed the rise and fall of the dinosaurs and the relatively recent arrival of humans on its big blue partner.

The Moon at first quarter phase, imaged with a 127mm refractor and digital camera.

Our lunar companion is instantly recognizable to anyone who has ever looked at the night sky. Measuring 3,476km (2,160 miles) across, the Moon measures around a quarter the diameter of the Earth – about the same breadth as the United States. The Moon is our only known natural satellite, a sphere of rock whose size is only exceeded by four other satellites in the Solar System. Because of the Moon's relatively large size, the Earth and Moon have often been called a 'double planet'.

Lunar origins and history

It's incredible to imagine that the Moon might have been produced in a one-in-a-million chance collision between a small planet and the young Earth, but this is currently the most widely accepted theory to account for the Moon's origin. Other theories include the idea that the Moon was flung off the Earth by rapid rotation, that the Moon is a captured planet or that the Moon formed out of a ring of debris left over from the Earth's formation, but none of these theories quite fits all the data we have about the Moon. The 'Big Whack' theory envisions a Mars-sized planet colliding with the Earth. The impactor's heavy core joined with the material of the Earth, while the lighter molten mantle material of the impactor was mixed with that of the Earth and flung out into space, where much of it coalesced in orbit to form the Moon.

Once the Moon had formed and its crust had solidified, the Moon experienced a phase of intense asteroid bombardment. Our own planet was similarly assaulted from by the debris left over from the formation of the Solar System, but it has been obliterated by the incessant action of plate tectonics deforming the crust, in combination with volcanic activity, erosion and sedimentation, processes that do their best to hide the topography of past epochs. But the Moon's crust has long been solid and immovable, and etched deep into its face can be seen the signs of both volcanic activity and asteroid impacts. Most of this activity took place billions of years ago, long before dinosaurs had gained a clawhold on the Earth. Yet we can clearly see many of these ancient features through binoculars and telescopes. Some really ancient craters, formed billions of years ago, are so well preserved that the uninformed observer might imagine that they had been formed in recent times.

With its relatively small mass and low gravity, the Moon never managed to hold on to a substantial atmosphere, nor did liquid water ever gush across its surface. No lunar life of any kind has ever appeared, and the Moon has remained utterly sterile since its formation, 4.6 billion years ago.

The Moon in space

At an average distance of 384,400km (238,800 miles), the Moon revolves around the Earth in a near-circular orbit once every 27.3 days and presents a disk about half a degree across – small enough to be covered by the tip of your little finger.

When furthest from the Earth, at apogee (*apo*: far, *gee*: Earth), the Moon can be as distant as 406,700km (252,700 miles); when closest, at perigee (*peri*: near), it can approach to 356,400km (221,400 miles). At perigee its angular diameter is almost 33.5 arcminutes, while at apogee it measures less than 29.5 arcminutes – this means that the perigee Moon has an apparent area almost one-third greater than the apogee Moon.

As it orbits, the Moon maintains one face turned towards the Earth, keeping most of its far side perpetually hidden from our view. A phenomenon known as libration – a rocking motion of the Moon's globe, too slow to be observable in real time – means that from the Earth a total of 59 percent of the Moon can be observed over a period of time. The remaining 41 percent of the lunar surface constitutes the side that is permanently hidden from the terrestrial observer.

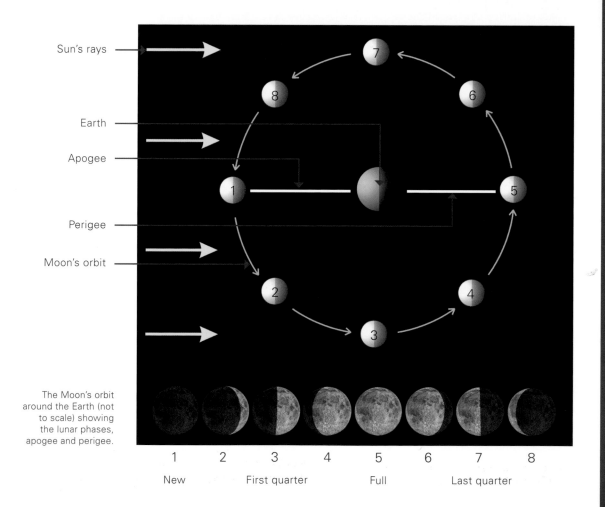

The Moon's orbit around the Earth (not to scale) showing the lunar phases, apogee and perigee.

1	2	3	4	5	6	7	8
New		First quarter		Full		Last quarter	

The Moon's phases

As the Moon orbits the Earth, its angle of illumination by the Sun slowly changes, and the shape of the illuminated portion of the Moon (its phase) changes from a thin crescent, through 'first quarter' (half Moon), to full Moon, 'last quarter' (half Moon), thin crescent, and back to new Moon. At its full phase the Moon is opposite the Sun in the sky and appears completely lit. At new Moon, the Moon lies between the Sun and the Earth and is not illuminated at all from our perspective. This cycle of lunar phases repeats itself every 29.5 days.

Moving west to east against the stars by its own diameter each hour, the Moon tracks east by about 13 degrees every day. The Moon's path lies close to the ecliptic, the course followed by the Sun through the Zodiacal constellations during the year. First-quarter Moon appears in a part of the sky where the Sun will be located three months later; last-quarter Moon appears where the Sun was three months earlier.

When the Moon is opposite to the Sun it appears fully illuminated, and is in the approximate position where the Sun will be in six months' time. Exact alignments of the Sun, Earth and Moon produce lunar eclipses as the full Moon passes through the Earth's shadow, but most of the time the Moon passes above or below the Earth's shadow without being eclipsed.

Around northern winter solstice the Sun is at its most southerly declination, while the full Moon rides high in Taurus or Gemini at midnight. At northern summer solstice, when the Sun is highest in the sky for the northern hemisphere, the full Moon at around this time appears very low down in Sagittarius, just peeking above the southern horizon at midnight. From the southern hemisphere the circumstances are exactly opposite.

The Moon's phases.

Earthshine

A phenomenon called 'Earthshine' – a faint illumination of the Moon's dark side – is most obvious when the Moon is a slender crescent set in a dark sky. This delightful spectacle, sometimes called 'the old Moon in the young Moon's arms', is caused by sunlight being reflected from the Earth onto the Moon's dark side. Through binoculars it's even possible to identify some of the larger surface features that are illuminated by the Earthshine alone.

The Moon illusion

When the full or near-full Moon is seen close to the horizon it often appears to assume unexpectedly large proportions. In reality, on any given night the Moon's apparent diameter is imperceptibly smaller when it is near the horizon because of our viewpoint on the Earth.

The Moon's apparent diameter always remains around half a degree, regardless of its location in the heavens. Nevertheless, the Moon illusion is powerful. It occurs mainly because we perceive the sky as being shaped like the interior of a flattened dome – an illusion reinforced by the perspective effect of clouds and other aerial objects seen during the daytime, which all scoot beneath this dome. Celestial objects appear attached to the imaginary dome's interior. When the near-horizon Moon is viewed we imagine that it is far away and must therefore be a large object to subtend that angle. When the Moon is above us we subconsciously imagine it to be closer to us and smaller in size. Exactly the same illusion occurs in our perception of the size of constellations. Castor and Pollux, for example (*see* The Starry Skies), appear more widely separated when they are near the horizon.

Earthshine illuminates the dark side of the young crescent Moon, imaged with an undriven digital SLR.

We perceive the Moon low on the horizon as being larger than when it is higher in the skies. The Moon illusion is a perception that is 'hard wired' into our brains.

Lunar eclipses and occultations

Sometimes the full Moon moves through the shadow cast by the Earth into space, and experiences an eclipse. On such occasions the Moon doesn't completely vanish from sight, even if it happens to plunge deep into the Earth's shadow, because a certain amount of sunlight is bent around the edge of the Earth by the atmosphere and reaches the Moon's surface. This bent light is however strongly red in colour, and as a result the totally eclipsed Moon appears reddish in colour. Its hue varies from eclipse to eclipse, ranging from bright orange to deep ruddy brown. A lunar eclipse can last for several hours from the point that it first enters the Earth's shadow to the moment it leaves the shadow. Lunar eclipses are best viewed through a pair of steadily held binoculars – the viewer gets the wonderful impression of the Moon floating in space at totality, with the darkened Moon set among a starfield that might normally be invisible at full Moon.

During its course through the skies, the Moon frequently passes in front of stars and hides them from view for a while. Known as occultations, these events can be fascinating to view. Often, a star will appear to switch off as the Moon's edge covers it – a dramatic event that never fails to astonish the observer by its sheer suddenness. Occasionally a planet is occulted by the Moon, and such events are eagerly anticipated, as the occultation of a large object such as Venus, Jupiter or Saturn can take many tens of seconds to complete; to have the Moon and a planet in the same high-magnification field of view provides a wonderful visual delight. Dates and times of bright lunar occultations are printed in a number of publications.

The sublime beauty of a total lunar eclipse, imaged with a 160mm refractor and digital SLR.

An occultation of Saturn by the Moon, observed by the author using a 100mm refractor.

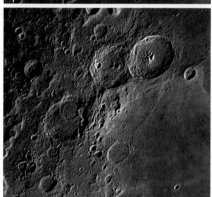

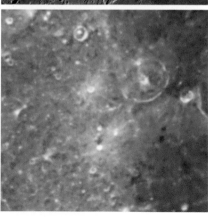

Sightseeing the Moon

Being so close to the Earth, the Moon appears large and bright, and the biggest of its surface features can be seen with the unaided eye alone. It's thrilling to view the Moon through binoculars. A number of very large dark flattish areas surrounded by mountain ranges are the most obvious features on the lunar near side. These are the maria (Latin: seas) – vast expanses of lava that flowed and solidified several billion years ago. Maria are generally circular in shape because they occupy the floors of huge asteroid impact basins. In contrast, the Moon's unobservable far side has very few smooth lava-flooded areas – most of it is exceedingly rough and pockmarked with craters of all sizes.

Craters abound on the Moon, especially in the crater-crowded southern uplands of the near side. It's estimated that the lunar surface is dotted with three trillion (three million million) craters larger than a metre in diameter and tens of thousands of these are large enough to be seen through a small telescope. The overwhelming majority of lunar craters were formed by the impact of asteroids, comets and meteoroids, and though most craters share the same origin, no two look exactly alike. Some ancient craters are highly eroded and are only visible at low angles of illumination. Other craters show magnificent structure in their walls and impressive mountains rise at their centres, while a number of large craters have been largely flooded with lava.

As the Moon's terminator (the line of sunrise or sunset) moves across the Moon's face, features near to it appear bright and prominent, and craters are filled with shadow. Seeing this, the viewer might imagine that the craters are incredibly deep hollows and the mountains are pointed and spire-like. But this is an illusion caused by the low angle of illumination. Further away from the terminator, features appear less severely sculpted; indeed, under a high afternoon Sun no shadows are cast, and some features disappear completely, totally blending into their surroundings.

Binoculars show hundreds of impact craters, and their forms vary depending on their age, size and how much other things (such as volcanic activity and overlying impacts) have altered their appearance since they were formed. The youngest craters are sharp and bright, and many are surrounded by prominent rays of ejected material that can stretch across the lunar surface for hundreds of kilometres.

A telescope will show much more on the lunar surface. Large and impressive mountain ranges rise at the borders of many of the maria, most notably around

Illumination changes our perception of the ruggedness of the lunar surface. Here, craters Theophilus, Cyrillus and Catharina are seen at sunrise, early morning and under a lunar midday Sun.

the Moon's largest, Mare Imbrium (Sea of Rains), whose edge is marked by the Jura Mountains, the Alps, Apennines and Carpathians. These majestic ranges were formed directly as a result of the mighty asteroid impact that produced the basin of Mare Imbrium.

Lunar mountain ranges rise to impressive heights above their surroundings – some peaks in the Apennines, for example, reach heights of more than 5,000m (16,400ft). It's fascinating to watch these mountains cast dark, broad jagged shadows onto the marial plain just after lunar sunrise or towards sunset. In reality, the Moon's mountains don't look like the pointed, sharply angled mountains of Earth's Alps or Himalayas; they have relatively gentle slopes of around 30 degrees and are rounded as a consequence of billions of years of constant sandblasting from the impact of tiny meteorites.

Some parts of the Moon's crust have experienced a pulling-apart movement, which caused the crust to crack along fault lines. A remarkable fault known as the Straight Wall forms a giant cliff face some 110km (68 miles) long in eastern Mare Nubium (Sea of Clouds). Long, curving fault valleys called arcuate rilles cut through the Moon's crust around the edges of some of the maria, and the floors of some individual craters are cut through by straight clefts called linear rilles. Most lunar rilles are rift valleys known as graben, formed when the crust between two close parallel faults sunk down. By far the largest of such features is the Alpine Valley, a giant crustal rift 160km (100 miles) long and 18km (11 miles) wide in places, which slices clean through the lunar Alps.

The northern border of Mare Imbrium, illuminated by an afternoon Sun. Nestled in the lunar Alps is the large, dark-floored crater Plato, and cutting across the mountain range is the Alpine Valley. Imaged with a 250mm reflector and astronomical CCD camera.

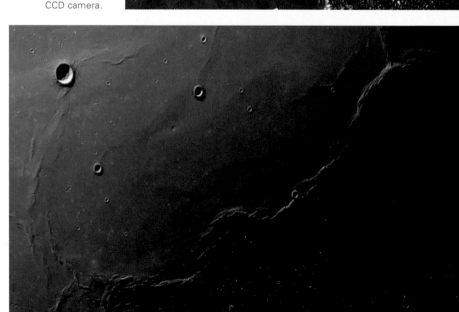

Dorsa Smirnov, a winding wrinkle ridge in Mare Serenitatis. Imaged with a 350mm SCT and astronomical CCD camera.

A number of rilles found in the maria take a winding path and appear like dried-up riverbeds. These features, known as sinuous rilles, are erosion features formed by fast-running rivers of molten lava. Schröter's Valley, the best example of a sinuous rille, can be found in Oceanus Procellarum (Ocean of Storms) near the brilliant ray crater Aristarchus. Beginning at a wide, distinctively shaped feature called the Cobra's Head, Schröter's Valley winds a 160km (100-mile) long path through the plain. In places it is more than 10km (6 miles) wide and 1,000m (3,280ft) deep and is visible through a small telescope when the Moon is two or three days before full.

Inside the maria are bays and ghost craters – old impact craters whose walls have been breached by lava flows and partly filled by them. Isolated mountains jutting out from many of the maria are remnants of the basin's inner walls that have avoided being completely submerged beneath the lava flows. Observation reveals low domes, rounded hills thought to be long-extinct lunar volcanoes, some complete with summit vents. Small valleys produced by flowing lava can be seen winding their way down the flanks of some domes. When illumination conditions are right, long, low ridges can be seen running across the maria; these are wrinkles formed after the lava flows had ceased and the material cooled and contracted.

Anomalous lunar activity

Transient lunar phenomena (TLP) are rarely observed localized coloured glows, obscurations and brief flashes. Most recorded TLP have been observed by amateurs so there's not a great deal of scientific evidence supporting most of the observations. It's possible that glows and obscurations are caused by lunar degassing and electrical activity. Flashes on the lunar surface have been recorded on video; coinciding with meteor showers, it is likely that they are caused by the impact of small meteoroids.

The waning gibbous Moon, (any phase between half-illuminated and full) a couple of days after full. Dark areas are lowland lava plains, while brighter areas are mountainous, more heavily cratered regions.

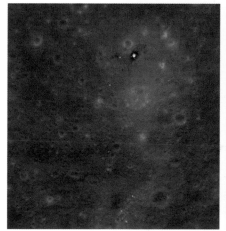

A comparison between the smoother lunar lowland plains (the Apollo 11 landing site in Mare Tranquillitatis) and cratered highlands (the Apollo 17 landing site in Taurus-Littrow), imaged by NASA's Lunar Reconnaissance Orbiter. The landing stages can be seen, as well as tracks left by the astronauts.

Both Mercury and Venus orbit the Sun closer than the Earth, and for that reason they are referred to as the inferior planets. They never appear to stray very far from the Sun, and the two planets can often be seen with the naked eye shining in the dawn or dusk skies.

Through a telescope both Mercury and Venus show a sequence of phases during each elongation from the Sun, in addition to changes in their apparent diameter. Following an inferior planet from its superior conjunction, when it lies directly on the far side of its orbit, hidden from view by the Sun's glare, the planet begins to edge east of the Sun. Once it has moved far enough away from the Sun to be seen against a reasonably dark evening sky, a telescope will show it to be a small gibbous disk (greater than a semicircle and less than a circle).

    The inferior planet's phase decreases as it moves further away from the Sun, until it reaches half phase, or dichotomy, at its greatest elongation east of the Sun. The planet then begins to move towards the Sun, its phase becoming a large crescent, until it is lost in the Sun's glare once more. Inferior conjunction is reached when Mercury or Venus passes between the Earth and the Sun.

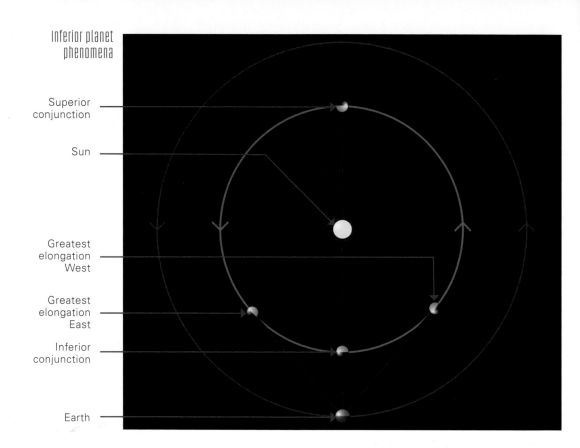

inferior planet phenomena

- Superior conjunction
- Sun
- Greatest elongation West
- Greatest elongation East
- Inferior conjunction
- Earth

Phenomena of the inferior planets. Eastern elongations take place in the evenings, western elongations in the mornings. At conjunction the Earth, inferior planets and Sun are roughly in line – inferior conjunction when the planet is on the Earth's side of the Sun, superior conjunction when on the far side of the Sun.

On most occasions the planet will pass some distance to the north or south of the Sun at inferior conjunction, but on those rare occasions when an inferior planet happens to pass exactly between the Earth and Sun, it appears to transit the Sun's disc and can be seen as a small circular silhouette. As the planet draws to the west of the Sun, it eventually becomes visible in the pre-dawn skies, when telescopes will reveal it to be a large crescent phase. The planet's phase gradually fills out, and its apparent diameter slowly decreases. Dichotomy is reached at its greatest elongation west of the Sun. As the planet draws closer to the Sun once more, it becomes a more gibbous phase, growing ever smaller, until it is lost in the Sun's glare. The sequence begins again at superior conjunction.

Not all morning and evening apparitions of Mercury and Venus are favourable. From temperate latitudes, the angle made by the ecliptic at the sunset or sunrise horizon varies from season to season. For observers in both hemispheres, eastern elongations of Mercury and Venus are most favourable during the spring, when the planets are highest above the western horizon as the Sun sets. Western elongations of the inferior planets are most favourably observed during the autumn (fall), when they are at their highest above the eastern horizon before sunrise.

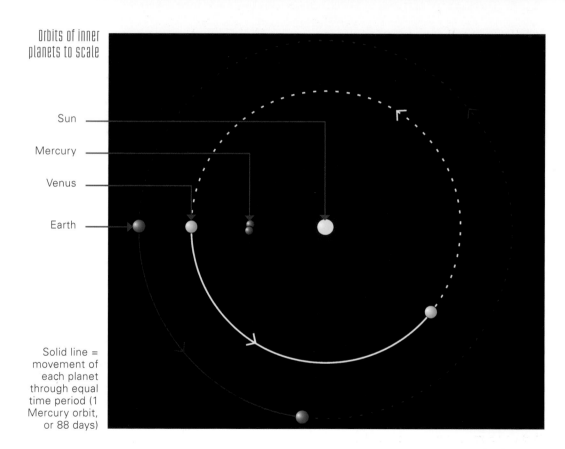

Orbits of inner planets to scale

Sun

Mercury

Venus

Earth

Solid line = movement of each planet through equal time period (1 Mercury orbit, or 88 days)

Both Venus and Mercury have shorter orbital periods than the Earth. Mercury completes one orbit in 88 days, while Venus takes 225 days. The Earth's orbital period is, of course, 365 days. This shows the relative motions of the planets in one Mercurian year of 88 days.

Mercury is the most elusive of the planets visible with the naked eye. It whizzes around the Sun once every 88 days, and is only favourably placed for viewing for a short period (around a week or two) when at its greatest angle from the Sun. City dwellers may find Mercury a tricky planet to spot because it always appears rather low down near the sunrise or sunset horizon. To be able to locate Mercury you need a reasonably clear horizon, and the planet must be high enough for its light to shine through the near-horizon atmospheric murk. Mercury is best placed for viewing from the northern hemisphere in the evening skies of spring or the morning skies of autumn (fall), when it is at its highest (around ten degrees) above the horizon, at least half an hour after sunset or before sunrise respectively.

Mercury can appear almost as bright as Sirius, the sky's brightest star, and it shines with a rosy hue. Through a telescope, this little world – a bare rock around twice the diameter of the Moon – doesn't show a great deal on its small disk, though its phase can be discerned at a high magnification. It was only when the Mariner 10 space probe imaged the planet more than 30 years ago that we first got to know that its surface is very heavily cratered, like the Moon's highlands.

Through a telescope Mercury presents a rather small disk. Its phase is noticeable at higher magnifications, but surface markings are subtle and elusive. Many observers choose to observe Mercury when it is high in the sky during full daylight, either by continuing to follow the planet after sunrise, or by using a telescope with a computerized go-to system to locate it. It is potentially dangerous to attempt to locate Mercury (or for that matter any other object near to the Sun) with a telescope during the daytime by sweeping the skies in a haphazard fashion – a brief magnified flash of sunlight on the eye is enough to permanently damage a person's vision.

Mercury, imaged by NASA's MESSENGER probe. The planet is heavily cratered, and some of the younger craters are surrounded by bright splashes of ejected material.

Venus – a deceptive beauty

Venus is a bright, conspicuous object in the evening or morning skies. Seen in a dark sky, the planet can be so dazzling that it is sometimes mistaken for a UFO by uninformed people – notably by US President Jimmy Carter, who filed an 'official' report on his so-called 'close encounter'!

Venus's orbit around the Sun takes a leisurely 225 days, which means that there can be a maximum of two elongations from the Sun each year. At its maximum, Venus reaches a respectable angle of 45 degrees from the Sun, so it can appear high in the sky for several hours after sunset or before sunrise. Venus is best placed for viewing in the evening skies of spring or the morning skies of autumn (fall) of either hemisphere, when it is at its highest above the sunset or sunrise horizon respectively.

Around the same size as the Earth, Venus is swathed in clouds so thick that its surface can never be seen through the telescope. It was once thought that conditions on Venus might be suitable for life to flourish, but that notion changed utterly when space probes went there and landed on its surface. Venus, named after the beautiful goddess of love, is the nearest planetary environment in the Solar System to old-fashioned ideas about hell! Clouds of sulphuric acid drift through an atmosphere of carbon dioxide. On Venus's surface the atmospheric pressure is higher than in a pressure cooker, and temperatures are far hotter than a kitchen oven.

Seen through a telescope, some Venusian cloud features are occasionally visible, but they are usually indistinct. However, the planet's phases are obvious, varying from a small gibbous disk to a sizeable crescent during each apparition.

Being an inferior planet, Venus sometimes passes directly between the Sun and the Earth, appearing as a black spot that travels across the Sun in a few hours. Transits of Venus are rare events, taking place in pairs eight years apart, separated by gaps of more than a century. The transits of June 2004 and 2012 won't be repeated until December 2117 and 2125.

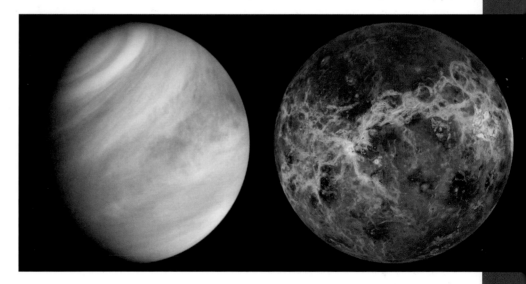

Venus, imaged by NASA's Mariner 10 (left) and Magellan (right). While Mariner's image shows only the planet's bright cloud tops, Magellan's radar image penetrates the thick atmosphere to reveal a strange volcanic surface.

Mars, Jupiter, Saturn, Uranus and Neptune are collectively known as the superior planets because they all orbit the Sun outside the orbit of the Earth. Conjunction with the Sun takes place when a superior planet is on the far side of the Sun from the Earth. Since the Earth is on a faster orbital circuit than the superior planets, they appear to edge west of the Sun after conjunction. As they clear the Sun's glare, the superior planets begin to peek out of the pre-dawn skies. At this stage the superior planets are at their smallest apparent diameter.

Jupiter always appears larger than 30 arcseconds across, and the apparent diameter of Saturn's disc always exceeds 18 arcseconds. Mars, however, only becomes visually interesting when it grows larger than 5 arcseconds in apparent diameter, when its ice caps and deserts can just about be discerned through a 100mm telescope. This occurs many months after Mars first becomes visible with the naked eye in the morning skies.

At opposition, each superior planet is located opposite the Sun in the sky, and it is due south at midnight. The superior planets are virtually 100 percent illuminated at opposition and are at their largest apparent diameter for that particular apparition. Since each planet has a slightly

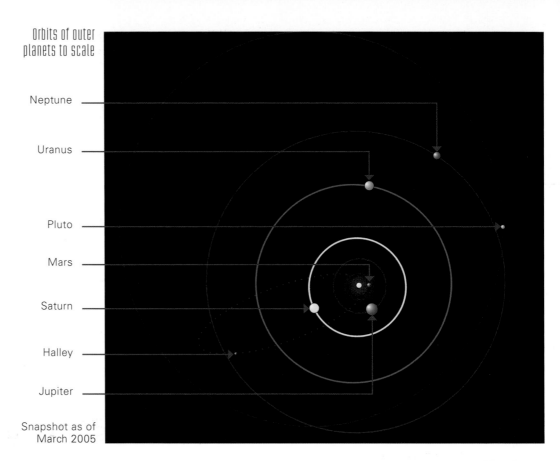

Orbits of outer
planets to scale

Neptune

Uranus

Pluto

Mars

Saturn

Halley

Jupiter

Snapshot as of
March 2005

The orbits of the five outer planets – Mars, Jupiter, Saturn, Uranus and Neptune – with the orbits of dwarf planet Pluto and Halley's Comet.

elliptical orbit, the distance between the Earth and the superior planet varies at opposition. Mars displays the greatest variation in opposition diameter, ranging from a minimum of around 15 arcseconds across at aphelic oppositions (when it is furthest from the Sun) to 25 arcseconds at perihelic oppositions (when it is nearest the Sun).

Of course, the slow drift of the superior planets to the west of the Sun during their apparitions is caused by the movement of the Earth around the Sun. Although the overall movement of the superior planets against the celestial background is slowly eastwards, a phenomenon called retrograde motion causes a superior planet to reverse direction among the stars for a while, performing a small loop or zigzag in the sky before proceeding eastwards, caused by our view of the planet from the Earth and our shifting line of sight.

As we draw further away from the planet after opposition, our line of sight begins to alter the apparent course of the planet; it appears to slow down and then recommences its slow eastward path. The size of the retrograde paths made against the background stars lessens with the distance of the planet. Minor planets, orbiting in the main asteroid belt between Mars and Jupiter, also display retrograde motion during the course of an apparition.

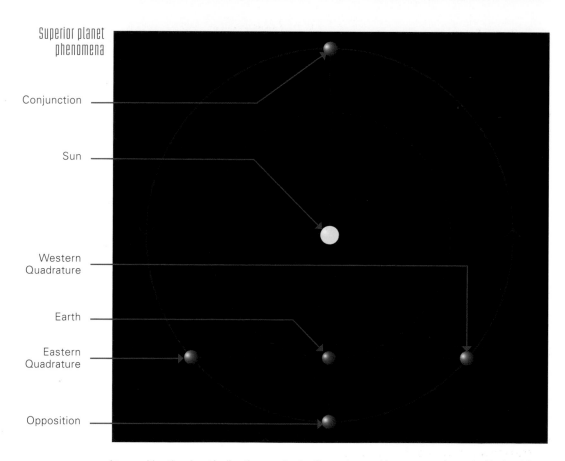

Superior planet phenomena

Conjunction

Sun

Western Quadrature

Earth

Eastern Quadrature

Opposition

At opposition the planet is directly opposite the Sun and around its nearest point to the Earth. while at conjunction it lies on the far side of the Sun and its furthest from Earth. At quadrature the planet-Sun-Earth angle is 90 degrees – western quadrature appearing in the morning, eastern quadrature in the evening.

Of all the planets in the Solar System, the red planet Mars has attracted the most attention from both science and popular culture because it has long been known to be similar to the Earth in many ways.

Orbiting further from the Sun than the Earth, Mars is just over half the Earth's size with a surface area roughly equal to that of the Earth's dry land. Its day is only 37 minutes longer than our own, and its axial tilt is similar to the Earth's, meaning that the planet experiences seasons. For centuries, astronomers have watched in fascination as Mars's polar ice caps vary in size with the seasons, suggesting that the ice melts when the planet warms up in the Martian summertime and refreezes during the winter. We now know that the caps are made up of both frozen carbon dioxide and water ice. Seasonal changes affect the dusky markings on Mars, too, as some areas appear to broaden and darken, while others fade. Yet the markings always appear to return to their original configuration, regardless of any temporary changes they might experience.

Mars has long been known to have an atmosphere, too, as occasional bright clouds and large dust storms have been observed through the telescope. Some dust storms have been so severe as to completely obscure all the markings on the planet.

Martians?

Being so like our own world, it's hardly surprising that many astronomers concluded that Mars was a likely abode of life. Some astronomers imagined that they could see networks of lines stretching across the Martian surface. It was suggested that these (non-existent) features might be canals constructed by intelligent beings to transport meltwater from the ice caps to irrigate the arid Martian deserts. Science fiction writers often depicted advanced civilizations on Mars, some of whom dared to attempt to conquer our own planet!

Science has yet to prove the existence of life on Mars, past or present. Space probes have imaged ancient dry riverbeds and lakes, in addition to detecting minerals that could only have formed in water, so water once flowed in great quantities across the planet's surface when it was warmer several billion years ago. Water ice exists at the Martian poles and buried in vast quantities beneath large swathes of its surface, and in some places it occasionally seeps out in liquid form,

Mars, imaged by NASA's Mars Global Surveyor. The north polar cap is visible, along with clouds over the planet's large extinct volcanoes.

producing small channels in steep sunlit slopes. It is not unreasonable to suspect that primitive microbial life may have developed on Mars – indeed, it might be there today, hidden in the planet's red soil.

Mars's surface

Space probes have settled many mysteries of the Red Planet. Mars is colder than the Earth, and it has a thin atmosphere of carbon dioxide. Humans could not exist there without a spacesuit. Bright clouds of ice crystals occasionally form, and the wind whips up anything from small dust devils to occasional seasonal dust storms, which obscure large swathes of the planet. The wind is responsible for the changing appearance of the dusky features on Mars, as the planet's dark surface material is temporarily hidden by dust or exposed as the dust is blown away.

Rocky deserts cover much of Mars, coloured red because they are literally rusty with the mineral iron oxide. Craters predominate in the planet's southern hemisphere, while large portions of the northern hemisphere consist of rolling plains.

An impressive collection of extinct volcanoes dominates the Tharsis region.

Of these Mount Olympus, the Solar System's biggest volcano, is 500km (310 miles) wide and three times higher than Mount Everest. Clouds over Mount Olympus can sometimes be seen through backyard telescopes. East of Tharsis lies the incredible Mariner Valley, a giant rift more than 3,000km (1,860 miles) long, 600km (370 miles) across and 8,000m (26,250ft) deep in places. Some dark parts along the valley floor can be seen through a telescope.

Observing Mars

Most of the time Mars is too small for much detail to be seen, but for a few months every couple of years the planet comes close enough for its ice caps and its dusky deserts – even bright cloud features on occasion – to be easily made out through a modest telescope.

Most of Mars's dark markings are located in the southern hemisphere. The wedge-shaped plain of Syrtis Major, along with Hellas, a bright area to its south, are familiar to all Mars observers. On the other side of the planet, the dark eye-shaped Solis Lacus and the broad dusky tongue of Mare Acidalium can appear striking. Mars's two satellites, Phobos and Deimos, are far too small and dim to be seen by casual backyard observers.

Mars, observed by the author with a 200mm SCT during close and far oppositions, showing how its apparent diameter varies with distance.

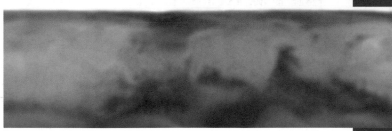

Map of Mars based on combined CCD images taken with a 222mm reflector in early 2010.

Jupiter – king of the planets

A dozen times broader than the Earth, Jupiter is the Solar System's biggest planet – a rapidly spinning ball of gas whose cloud features are fascinating to view through backyard telescopes. Jupiter is more voluminous than all the Solar System's other planets, satellites and asteroids put together. Jupiter has a fantastically fast rate of spin, revolving once on its axis in less than ten hours. Since the planet is made up of gas (mainly hydrogen and helium) its rapid spin produces marked centrifugal bulging at its equator. In addition to being a fascinating planet in itself, Jupiter's other attraction is the dance of its four largest moons – Io, Europa, Ganymede and Callisto – which are bright and easy to spot through binoculars.

Dynamic atmosphere

Jupiter has no visible solid surface. Its upper atmosphere is in constant turmoil, so no permanent features exist on the planet. Banding of the planet's dark belts and light zones, produced by the planet's rapid spin, can be seen through a small telescope. The belts and zones vary in intensity from year to year, but the most prominent are usually the North and South Equatorial Belts. Features within the cloud belts and zones change from week to week, as spots, ovals and festoons develop, drift in longitude, interact with one another and fade away. Features remain within their own belt or zone, and they never drift in latitude.

Astronomers have kept a constant watch on developments on Jupiter for more than a century. The nearest thing that Jupiter has to a 'permanent' feature is the giant anticyclone known as the Great Red Spot. Located in the South Tropical Zone since at least the mid-19th century, the Great Red Spot varies in intensity from year to year, ranging from a barely discernable grey smudge to a sharply defined brick-red oval.

Jupiter and its four large moons (not to scale), from images by NASA's Galileo probe. The Great Red Spot – a long-lived storm bigger than Earth, first observed in the 19th century – is visible.

Observing Jupiter

Jupiter orbits the Sun every 12 years, and as a result moves slowly among the Zodiacal constellations. It appears a brilliant white star, second only to Venus (and occasionally Mars at its brightest), so it isn't difficult to locate with the unaided eye.

Steadily held binoculars will reveal the four largest satellites of Jupiter – Io, Europa, Ganymede and Callisto – as four bright star-like points of light. Some people have claimed to be able to see one or more of them with the unaided eye.

Although Jupiter appears as a small flattened disk through a small telescope at low magnification, a high magnification using at least a 100mm telescope is required to see much detail within the planet's dark belts and bright zones. Features appear near Jupiter's western edge, and rotation carries them to the central meridian – the imaginary line connecting a planet's poles – in just a couple of hours. Within a few more hours rotation has transported them out of sight beyond the disk's edge.

Usually there's a good deal of activity going on in the equatorial regions, with festoons whipping off the south edge of the North Equatorial Belt, skirting around bright areas within the Equatorial Zone itself. Nestled in the South Tropical Zone, producing a marked indentation in the south edge of the South Equatorial Belt, the Great Red Spot is usually visible when it is on the hemisphere facing the Earth, though its colouration may not be glaringly obvious. Small spots and ovals are often visible within Jupiter's atmosphere, and it's enjoyable to make a sketch of the planet's appearance and trace the apparent movement of these features within the Jovian atmosphere over a period of weeks or months.

Jupiter's four bright satellites – known as the Galilean moons – are fascinating to follow through a telescope. Io, Ganymede and Callisto

are larger than our own Moon, and Europa is only slightly smaller. These planet-sized worlds can just about be discerned as tiny disks through a 100mm telescope, although no detail can be seen on them. As they orbit Jupiter in the plane of its equator, the Galileans occasionally appear to move directly across the face of Jupiter, preceded by (or trailing) the round black shadows they cast on the Jovian disk. Frequently eclipsed by the shadow of Jupiter, the satellites play a continual game of hide and seek as they peek out from, or slip behind Jupiter's limb. Sometimes, mutual occultations and eclipses between Jupiter's moons take place.

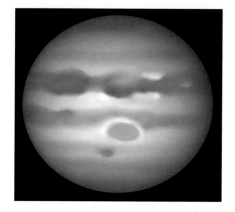

Jupiter, observed by the author using a 200mm SCT.

Jupiter, imaged with a 250mm reflector and astronomical CCD camera. Io is also visible.

Saturn's sheer beauty invariably surprises and awes the first-time telescopic viewer. With its wonderful ring system and array of moons, Saturn never fails to provide a great visual delight.

Saturn orbits the Sun way beyond Jupiter, completing a circuit in a little more than 29 years. Like Jupiter, Saturn is a giant, rapidly spinning ball of gas (93 percent hydrogen, the lightest gas), which is noticeably flattened at the poles. Far less dense than Jupiter – considerably lighter than water, in fact – the planet would float if you could find a big enough pool of water in which to place it.

Saturn's atmosphere displays dusky bands and brighter zones parallel to the equator, but they are less pronounced than those seen on Jupiter. There's much less obvious activity within Saturn's atmosphere too, and localized features that appear distinct enough to be seen through the telescope, such as dark spots or bright ovals, are rather rare. Large bright spots billowed up in Saturn's atmosphere in 1933, 1960 and 1990, but these were all short-lived and faded after a few weeks. In 2010 a much longer-lived atmospheric disturbance arose in the northern hemisphere, which experienced an unprecedented second eruption in 2011.

Saturn's remarkable rings

Through their small, somewhat imperfect telescopes, early observers didn't know how to interpret the strange features that they saw poking out from either side of Saturn. In 1656, Christian Huygens rightly suggested that Saturn was surrounded by 'a flat ring nowhere touching the planet.' The ring system is exactly in line with the planet's equator.

Saturn is titled to the plane of its orbit around the Sun, and our view of the rings changes year after year. From being at their widest, the rings appear close to a thin line in around seven years; after another seven years, the rings once more appear wide open, but with Saturn's other pole presented to us.

Saturn's ring system can be split into three main components. Ring A, the outermost ring, is slightly darker than Ring B, the middle main ring, and they are separated by a narrow gap called the Cassini Division. Both the A and B rings appear uniform and opaque, and they cast a dark shadow onto the planet below. The innermost ring, Ring C, is difficult to observe because it is very faint and somewhat transparent – qualities that inspired its unofficial name of the 'Crepe Ring'.

Saturn and its large moon Titan (not to scale), from images by NASA's Cassini probe.

Billions of individual particles, ranging from dust grains to chunky boulders, make up Saturn's rings. Space probes have shown that the main rings actually comprise hundreds of individual narrow ringlets of varying opacity. NASA's Cassini probe, which entered into orbit around Saturn in 2004, has returned magnificently detailed images showing how the gravitational pull of small satellites shepherds the ring particles, influencing the appearance of the rings.

Saturnian moons

Saturn has the most known satellites of any planet in the Solar System – a total of 62 are known at present, of which 53 have been given official names, but only a handful of them are visible through backyard telescopes.

Titan, Saturn's largest satellite, is larger than our own Moon, and hangs on to a substantial atmosphere. In 2005 the Huygens probe landed on Titan, returning detailed images of one of the most fascinating worlds in the Solar System, a world with outlandish weather, wind, clouds, rain,

rivers and lakes of methane and ethane. Titan is easy to spot through a small telescope. Four more of Saturn's larger satellites may be glimpsed through a 100mm telescope, namely Rhea, Tethys, Dione and Iapetus.

Observing Saturn

Saturn shines as a bright yellowish star, which is easy to identify in the night sky. Binoculars won't usually show the rings clearly, but Titan can typically be seen. The ringed planet is a truly beautiful sight through any telescope. At a low magnification, the planet and its satellites can be seen in the same field of view. Steady seeing is required for a really good high-magnification view of Saturn. In addition to a gap in the rings known as the Cassini Division, visible through small telescopes under good conditions, a much narrower gap called the Encke Division can sometimes be made out near the outer edge of Ring A. Depending on the angle at which we view the planet, the shadow of the rings on the globe and the globe's shadow on the rings can clearly be seen.

Saturn's changing tilt between 2006 and 2011, imaged with a 222mm reflector and astronomical CCD camera.

Uranus was discovered in 1781 by William Herschel, who noticed it as a faint object through his telescope when scanning the constellation of Gemini. At first, Herschel thought that the new object might be an incoming comet, but before long its orbit was calculated and it was shown to circle the Sun at a vast distance beyond Saturn – around 20 times the distance of the Earth from the Sun, and so far away one orbit takes it more than 84 years to complete. Uranus is another giant ball of gas, with a diameter more than four times that of the Earth.

If you know exactly where to look, Uranus can just about be seen with a keen naked eye if the sky is clear and dark. It's easy to spot in binoculars, but to be able to discern its disk a high magnification is required. Through a high-power eyepiece, Uranus appears as a tiny featureless disk with a pale greenish hue. Little of interest was revealed by the Voyager 2 space probe as it flew by the planet in February 1986, except for a dusky polar region and a few vague bands and spots. It wasn't the planet itself that proved to be the highlight of Voyager 2's flyby through the Uranian system, but its amazing retinue of small satellites, all of which are too faint to be seen through backyard telescopes. Uranus also has a ring system, but this is far less spectacular than that of Saturn, and so faint that it can only be seen on images taken through big telescopes.

Discovered in 1846 following a concerted telescopic search, Neptune is the smallest of the Solar System's gas giants. With an orbital period of 164 years, Neptune lies at the staggering distance of 4.5 billion kilometres from the Sun – so distant that sunlight takes about four hours to reach it. Despite its immense distance, Neptune's atmosphere is surprisingly active, as revealed when Voyager 2 sped past the planet in August 1989. The strongest winds in the Solar System were measured

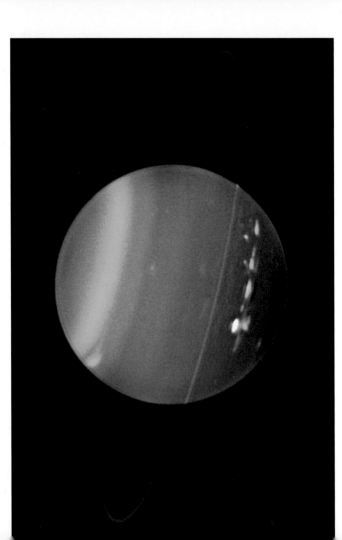

Uranus, imaged by the Keck Observatory.

to gust here at more than 2,000km/hour (1,243mph), and Voyager imaged a number of prominent atmospheric features, including a bright cloud dubbed 'the Scooter' and a 'Great Dark Spot' (like Jupiter's Great Red Spot), which was larger than the Earth. By 1994 the Hubble Space Telescope showed that these features had faded away, but similar ones elsewhere on the planet had sprung up.

Because of its faintness, Neptune is more challenging to locate than Uranus. Too dim to be seen with the unaided eye, it can be spotted as a star-like point through binoculars. Neptune's disk is tiny, and to resolve it requires at least a 100mm telescope with a minimum magnification of 200x. Observers might just be able to discern a slight bluish tinge – the result of methane gas absorbing the red wavelengths of sunlight falling upon it. Needless to say, obvious features on Neptune are not visible through amateur telescopes.

Beyond the eight officially recognised planets are numerous sizeable ice worlds. Pluto, the most famous of these, was discovered by the sharp-eyed Clyde Tombaugh on a photographic plate in 1930. Just two-thirds the size of our own Moon, images of Pluto taken by the Hubble Space Telescope have revealed a vague patchwork of light and dark areas, but what these might be is anyone's guess until NASA's New Horizons probe flies by in 2015. Most amateur astronomers have never observed Pluto – it is far too faint to be seen in anything smaller than a 250mm telescope, and even when seen through a really large telescope at maximum magnification it just looks like a faint star. Still, the thrill of actually seeing Pluto captivates many stargazers, worth every effort made in the hunt.

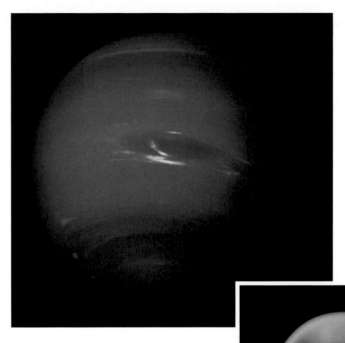

Neptune, imaged by NASA's Voyager 2 probe.

Pluto in true colour, imaged by the Hubble Space Telescope.

Comets and asteroids have had a bad press over the ages. In ancient times, people imagined that comets were harbingers of doom. More recently, movies have portrayed these cosmic itinerants as potential threats to human civilization.

Comets – celestial ghosts

When far away from the Sun, chilling out in the interplanetary deep freeze, a comet is a pretty unimpressive sight – a solid ball of ices mixed with dirt and rock measuring just a few kilometres across. Known as the nucleus, this 'dirty snowball' is made up of material left over from the formation of the Solar System.

As it approaches the Sun, the outside of the nucleus warms up, and the ices sublimate, turning into gas. Streaming off the nucleus, the gas carries with it grains of dust, forming a coma (Latin: hair), which envelops the nucleus in a haze many tens of thousands of kilometres across. Pulled by the Sun's gravity, the comet speeds up as it heads through the inner Solar System. Often a comet will develop a prominent tail which comprises two distinct components, one made up of gases and the other made up of dust. Gases are blown by the solar wind (energetic particles streaming out from the Sun), forming a straight tail that points directly away from the Sun. Countless billions of dust grains liberated from the nucleus form another tail. Pushed away from the nucleus by radiation pressure from the Sun, the dust grains follow their own orbit around the Sun in the wake of a comet, and as a result follow the comet's curving path through space. Reflecting sunlight, the curving dust tail is usually much brighter than the gas tail, and often shows considerable structure through binoculars. Although a bright comet's tail looks substantial, and can stretch for many tens of millions of kilometres, it is all cosmic showmanship. If you sampled

Comet C/2009 P1 Garradd passes in front of globular cluster M71 in August 2011.

the space near a comet you'd probably find that it was a better vacuum than can be produced in most laboratories on Earth. It has been said that a comet is the nearest thing to nothing that can be something.

A number of comets follow well-known orbits within the Solar System, the most famous being Halley's Comet, whose orbit takes it from the frigid realms beyond Neptune to the inner Solar System every 76 years. Some comets, such as 1997s spectacular Hale-Bopp, have orbital periods of many thousands of years. Another type of comet has never been warmed by the Sun before, having spent all its life at the very limits of the Sun's influence in a region known as the Oort Cloud. Thought to be a vast shell made up of billions of cometary nuclei, the Oort Cloud extends halfway to the nearest stars.

OBSERVING COMETS

Most newly discovered comets never develop into anything more spectacular than dim fuzzballs that can only be glimpsed through binoculars or telescopes. Now and again – say, once or twice a decade – a comet bright enough to be viewed with the unaided eye cuts a swathe through the heavens. Comet Hale-Bopp delighted stargazers all around the world, but really spectacular comets like these are rare, with perhaps only a handful coming into view every century.

Binoculars are ideal for observing comets, although sometimes the field of view won't be wide enough take in all of a bright comet's tail. Binoculars also deliver more colour to the eyes than a telescope eyepiece, and components of some tails can appear distinctly blue, red or green. At high magnification through a telescope the comet's tiny nucleus appears as a bright dot, sometimes surrounded by intricate structure such as jets, arcs and concentric shells of reflective dust.

Comet 103P/Hartley 2 passes near NGC 281 (the Pac Man Nebula) in October 2010. Imaged with a 105mm refractor and astronomical CCD camera.

Asteroids – vermin of the skies

When Han Solo piloted the Millennium Falcon through the Hoth asteroid belt in the movie *The Empire Strikes Back*, he had to concentrate hard to avoid colliding with thousands of giant space rocks tumbling across his path. When real spacecraft such as Voyager 2 sped through the Sun's main asteroid belt between Mars and Jupiter, mission controllers had no concerns that their billion-dollar probes would accidentally bump into anything. While there are hundreds of thousands of asteroids (known by astronomers as minor planets) within the main asteroid belt, they are spread throughout an enormous volume of space. If you were standing on an asteroid, the nearest asteroid would likely be a distant, dim star-like point of light.

None of the minor planets is bright enough to be seen easily with the unaided eye, and the first asteroid discovery had to wait until nearly two centuries had elapsed after the invention of the telescope. On 1 January 1801 – the first day of the 19th century – Giuseppe Piazzi discovered the first asteroid, which he later named Ceres. Ceres is far too small for its disk to be adequately resolved at the telescope eyepiece, but the methodical astronomer charted the object's slow movement among the stars of Taurus. Piazzi thought that the object might be an incoming comet, but after a while it became clear that the new object was a minor planet orbiting the Sun between Mars and Jupiter.

Reasoning that there might be yet more minor planets to be discovered, a group of European astronomers organized themselves into the Celestial Police to hunt them down. Minor planet Pallas was discovered in 1802, Vesta in 1804, and Juno was found in 1807. By the end of the 19th century several hundred minor planets were known. During the 20th century, so many asteroids were being discovered on photographic plates that they were sometimes rather unkindly referred to as the vermin of the skies. At the beginning of the 21st century, hundreds of thousands of minor planets have been identified. Of those orbiting in the main asteroid belt, Ceres is the largest, with a diameter of nearly 1,000km (621 miles); in 2006 it was designated a dwarf planet, a definition shared by the more distant icy worlds of Pluto, Haumea, Eris and Makemake. No fewer than 26 minor planets are in excess of 200km (124 miles) across. Yet even if all the Solar System's known asteroids were gathered together in one lump they would only form an object around half the size of our own Moon.

In addition to the minor planets within the main asteroid belt, there are other distinct groups of asteroids elsewhere in the Solar System. An interesting group called the Trojan asteroids are clustered at points far preceding and following Jupiter in its orbit around the Sun, locked there in a gravitational resonance. Of more concern to the news media are asteroids whose orbit takes them close to that of the Earth, including potentially hazardous asteroids – objects that could possibly collide with us at some point in the distant future.

Features on Ceres have been resolved by the Hubble Space Telescope, including what is thought to be a large impact crater that has exposed the dark mantle beneath its crust. A number of asteroids have been viewed up close by space probes. Each one has been shown to be battered and dented by previous impacts, both large and small. En route to Jupiter, the Galileo probe flew past Gaspra in October 1991, followed by Ida in August 1993. NASA's NEAR Shoemaker probe flew by asteroid Mathilde in June 1997 and went on to orbit asteroid Eros in 2000, finally making its way down to the asteroid's cratered, rubble-strewn surface in July 2000 after having orbited it for several weeks. In 2011 NASA's Dawn space probe went into orbit around Vesta, securing amazing images of a surface that is thought to have been covered with lava in its early history and later pummelled by smaller asteroids and meteoroids.

It is thought that many asteroids in the main belt were once part of larger parent bodies, which grew hot inside and developed a core, mantle and crust; these formed early in the history of the Solar System and were somehow broken up into smaller fragments. Most of the meteorites found on the Earth are pieces of asteroids – some of them undoubtedly originate from Vesta – and by studying their composition scientists can tell what materials make up the original parent objects.

OBSERVING ASTEROIDS

Although the minor planet Vesta can sometimes reach the border of naked-eye visibility, most asteroids are too faint to be seen without optical aid. Binoculars will reveal around a dozen of the larger main belt asteroids at their brightest, when they are opposite to the Sun in the sky. To be able to identify an asteroid, a good chart showing its position among the stars is required. Regular observation over a period of days and weeks will reveal the asteroid's path, and it can be a rewarding exercise to produce a plot based on your own observations.

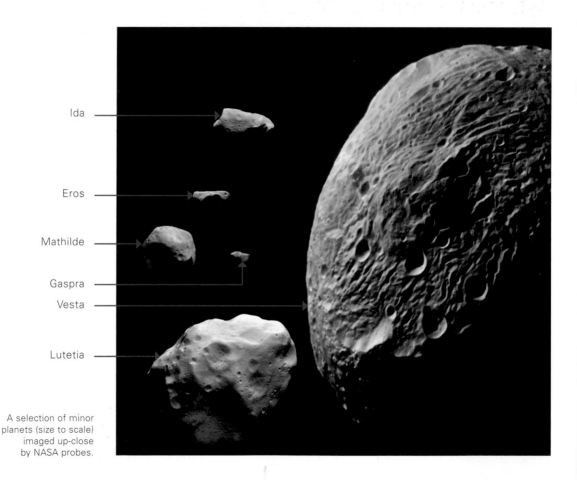

Ida

Eros

Mathilde

Gaspra

Vesta

Lutetia

A selection of minor planets (size to scale) imaged up-close by NASA probes.

AFTERWORD: LIGHT POLLUTION

Wherever you live, each clear night there's no shortage of opportunities to observe a variety of objects and phenomena in the night skies, whether you're using the unaided eye, binoculars, or a telescope of any sort, small or large. But it is impossible to enjoy the night skies fully when observing under the glare of direct lighting from neighbouring properties because the eyes' pupils are unable to dilate to their largest, as they do under dark conditions.

Bright lights drown out all but the brightest celestial objects. Here, even the full Moon struggles to compete with a sodium streetlight.

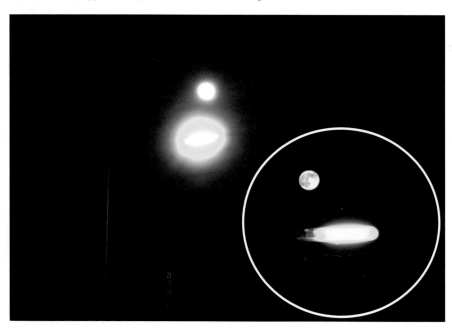

While large-scale urban light pollution may not be visible directly, it illuminates particles high in the sky, producing a ghastly glow that prevents faint objects from being seen.

As a whole, Solar System observation is far less hampered by the detrimental effects of light pollution than deep-sky observation, be it the type inflicted by the bright lights of neighbouring properties or the streetlights, industrial and commercial miasma blanketing whole urban regions. Whereas Solar System objects like the Moon and planets are bright enough to enjoy from virtually any location, nebulae and galaxies are very faint, low-contrast objects requiring a dark site and dark regional skies to appreciate fully. People living in urban areas rarely get the chance to see the wonderful band of the Milky Way, let alone explore its treasures and peer deeper into intergalactic space.

Virtually everyone is now affected by light pollution, destroying our view of the night sky; fewer than one in ten people in the developed world can see the true splendour of a starry sky. Light pollution is a visible and needless waste of resources, and the energy spent in illuminating the night sky is contributing to global warming. Around 1000th of the developed world's gross national product is wasted in electricity each year due to poorly designed or wastefully desired lighting. Light pollution doesn't only impact on stargazers. Poorly placed lighting has also resulted in disturbing people's enjoyment of their neighbourhoods, has a detrimental effect on wildlife and has caused medical health problems. In some cases light pollution has been the direct cause of human fatalities on the streets due to glare temporarily blinding motorists.

It is claimed that brightly illuminated homes, car parks and the like, reduces crime. There is no evidence to support this – in fact, lighting is very handy for criminal activity as it clearly illuminates a property's points of entry, gives owners a false sense of security and a lowering of vigilance; moreover, bright lights serve to mask criminal activity because the glare prevents them from being seen. For example, in some areas police have reported that night crime reduced dramatically after the street lighting was turned off overnight.

The idea that an area needs to endure daylight conditions 24 hours of the day, 365 days a year, should be abhorrent to anyone who appreciates the beauty of nature. Thanks to campaigning groups set up around the globe, darker skies are firmly on the agenda. Bob Mizon, Coordinator of the Campaign for Dark Skies (CfDS), has worked tirelessly in the UK since in 1989 to get the message across to individuals, companies and government agencies. Because astronomers are among the most sensitive to the effects of light pollution, the CfDS is a sub-section of the British Astronomical Association, although its membership is made up of a wide range of people, from lighting engineers to astrophysicists. The Society for Popular Astronomy in the UK also has its own light pollution advisory service with links to the CfDS. Nobody in the campaign advocates a return to the Dark Ages. Instead, Mizon states that the campaign's aim is the right amount of light only where needed. It's a point that few could reasonably argue with, but one that still needs to be taken more seriously.

In recognition of his work, Bob Mizon was awarded an MBE (Member of the Order of the British Empire) by Her Majesty Queen Elizabeth II in her 2010 Birthday Honours List 'for voluntary service to Astronomy and to the Environment'.

A group with similar aims, the International Dark-Sky Association, was established in the United States in 1988, their stated mission 'to preserve and protect the nighttime environment and our heritage of dark skies through quality outdoor lighting'.

Many government agencies and local governments are now beginning to turn off unnecessary on-all-night street lighting. It's true that the measures are in some way prompted by budgetary savings, and this has had the unfortunate effect of stirring misinformed people who want to keep the lights switched on for 'safety's sake' – yet this argument is fallacious. Better-designed home, commercial and street lighting can direct light down to where it is needed and not directly into the sky, and the CfDS has fostered links with engineers to stress this point. Small steps are leading to positive results that enhance everyone's lives.

The Milky Way running through Crux, Carina and Vela, imaged with an undriven digital camera for 30 seconds from a dark garden in New Zealand. The splendour of the scene is diminished by light pollution.

GLOSSARY

Altitude
The angle of an object above the observer's horizon. An object on the horizon has an altitude of 0 degrees, while at the zenith its altitude is 90 degrees.

Aperture
The diameter of a telescope's objective lens or primary mirror, usually measured in centimetres or metres.

Arcminute
A measure of angle equal to 1/60 of a degree.

Arcsecond
A measure of angle equal to 1/60 of an arcminute or 1/3600 of a degree.

Amateur astronomer
A person who enjoys observing astronomical objects and phenomena for sheer enjoyment.

Asteroid
A large chunk of rock orbiting the Sun, ranging from a few tens of metres to a few hundred kilometres in diameter. Also called a **minor planet**.

Asteroid belt
A zone of the Solar System containing a large number of asteroids. The main asteroid belt lies between the orbits of Mars and Jupiter.

Astronomical Unit (AU)
The average distance from the Earth to the Sun – around 150 million kilometres (93 million miles).

Astronomy
The science-based study of celestial objects and phenomena.

Aurora
The light show produced when energetic particles from the Sun get trapped in the Earth's magnetic field and collide with molecules in the Earth's upper atmosphere. The Aurora Borealis takes place in the northern hemisphere; its mirror image, the Aurora Australis, takes place in the southern hemisphere.

Big Bang
The colossal explosion at the beginning of time that created the Universe and everything in it. It took place around 13.7 billion years ago.

Binoculars
An optical instrument consisting of a pair of small parallel refracting telescopes, allowing both eyes to view simultaneously.

Black hole
A relatively small region of collapsed space-time with such a gravity field that nothing – not even light – can escape.

CCD
Charge-coupled device – a light sensitive electronic chip used for digital astrophotography.

Celestial sphere
From our perspective on the Earth's surface, the stars appear attached to the inside of a vast, all-encompassing sphere. The Earth's poles point directly to the celestial poles and the celestial equator is in the same plane as the Earth's equator. The celestial sphere appears to rotate around us, from east to west, as the Earth spins on its axis.

Comet
A city-sized chunk of ice and rock that heats up on entering the inner Solar System, emitting gas and dust which forms a coma, and perhaps developing a long tail of gas and dust.

Conjunction
Two or more close Solar System objects sharing the same Right Ascension (RA).

Constellation
A precisely defined region of the sky created to enhance our familiarity with the heavens. There are 88 recognized constellations (*see* How to Use This Book), some of which date back to remote antiquity.

Core
The central region of a star or large planet – usually very hot and under extreme pressure. Many minor planets are chips from larger objects and may not have a well-defined core.

Dark energy
A currently unknown force which is responsible for accelerating the expansion of the Universe.

Dark matter
A currently unseen form of matter that

makes up 90 percent of the mass of the Universe, known to exist because its gravity influences the motion of galaxies.

Degree
1/360 of a circle. The Sun and Moon are about half a degree across.

Double star
A pair of stars that appear close together in the sky. Some are binary systems orbiting each other; others, produced by line-of-sight perspective, are known as optical doubles. Systems of three or more stars are called multiple stars.

Eclipse
A phenomenon caused when a celestial object passes in front of, or through the shadow of, another celestial object. The Moon sometimes eclipses the Sun; the Moon itself is sometimes eclipsed by the shadow of the Earth.

Ecliptic
The line traced on the celestial sphere by the Sun throughout the year, corresponding with the Earth's orbit around the Sun. The planets orbit in planes roughly the same as the ecliptic.

Electromagnetic radiation
All energy in the electromagnetic spectrum – from short wavelength gamma rays to long wavelength radio waves – is propagated through space at the speed of light by vibrating electrical and magnetic disturbances. Visible light is a form of electromagnetic radiation.

Elongation
The apparent angular distance of an object from the Sun, measured between 0 to 180 degrees east or west of the Sun. For example, the first quarter Moon has an eastern elongation of 90 degrees; Venus has a maximum possible elongation of 47 degrees.

Exoplanet
A planet in orbit around a distant star. Hundreds are currently known.

Eyepiece
A lens inserted into a telescope that magnifies and focuses light into the eye.

Fireball
A very bright meteor caused by a large **meteoroid's** passage through the Earth's atmosphere.

Galactic cluster
A group of galaxies held together by their mutual gravitational attraction. Galactic clusters themselves belong to even larger superclusters – the largest structures in the Universe.

Galaxy
A large-scale agglomeration of matter, in which may be contained hundreds of billions of stars, held together by gravity and usually centred around a massive hub of stars at the centre of which lurks a supermassive black hole. Galaxies come in a variety of forms, some spiral, some elliptical, some irregular in shape.

Galilean moons
The four bright satellites of Jupiter discovered by Galileo in January 1610, which have been subsequently named Io, Europa, Ganymede and Callisto.

Gas giant
A very large planet which is composed largely of gas – mainly hydrogen and helium. Jupiter, Saturn, Uranus and Neptune are the Solar System's gas giants. They have no solid surface.

Geocentric
Having the Earth at its centre. The geocentric theory of the Universe postulated that everything in the cosmos revolved around the Earth.

Globular cluster
A collection of hundreds of thousands of individual stars, all held together in a vast sphere by their mutual gravity.

Heliocentric
Having the Sun at its centre. The heliocentric theory postulated that the Earth and planets revolved around the Sun.

Hertzsprung-Russell diagram
A graph that plots the luminosity of stars (y axis) against their surface temperature (x axis), devised independently by two astronomers: Ejnar Hertsprung and Henry Russell.

Light year
The distance travelled by light in one year. At a velocity of 300,000km per second (186,400 miles per second), light travels around 10 trillion kilometres (6.2 trillion miles) in a year.

Local Group
The cluster of galaxies to which our own Milky Way belongs. The nearby Andromeda Galaxy is its largest member.

Magnitude
The perceived brightness of a celestial object is called its **apparent magnitude**. The brightest star, Sirius, is magnitude -2.5, while the faintest stars visible with the unaided eye from beneath a dark sky are around magnitude 6. A star's real brightness, called **absolute magnitude**, takes into account the object's distance.

Main Sequence
Stars on the Main Sequence (of the **Hertzsprung-Russell diagram**) are ones that derive their energy from nuclear fusion of hydrogen into helium. Ninety percent of a star's lifetime is spent on the Main Sequence. The Sun is a Main Sequence star and is halfway through its lifetime of 10 billion years.

Meteor
A flash of light caused when a **meteoroid** burns up on entering the upper atmosphere.

Meteoroid
A small lump of rock in space. If it survives all the way down to the Earth's surface it is called a **meteorite**.

Milky Way
The name given to our home Galaxy. The distant stars in the galactic plane can be seen in a misty band encircling the sky.

Moon
The Moon is the Earth's only natural satellite. Satellites around other planets are also referred to as **moons**.

Nebula
A cloud of interstellar dust and gas. It may shine by reflecting the light from nearby stars, or by emitting its own light. **Dark nebulae** appear silhouetted against a brighter background. Old stars are sometimes surrounded by **planetary nebulae**, well-defined shells of puffed-off gas.

Neutron star
At the end of a massive star's life the star explodes as a **supernova** and the remnant core can form a fast-spinning neutron star composed mainly of neutrons and with a size approximately 10km (6 miles) in diameter. The density of a neutron star is similar to that of an atomic nucleus, so a thimble full of neutron star material will have a mass of a billion metric tons.

Nuclear fusion
Under high temperatures and pressures, energetic collisions between atomic nuclei form heavier nuclei, and a large amount of energy is released. The fusion process powers the Sun and other stars.

Parallax
The change in an object's apparent position with respect to more distant objects caused by a change in viewing angle. Nearby stars exhibit a measurable parallax, allowing their distance to be ascertained.

Red shift
The light from rapidly receding objects, such as distant galaxies, is stretched out into longer wavelengths, towards the red end of the spectrum. Red shift increases proportionately with the distance of galaxies.

Planet
A large non-stellar object in orbit around a star. The Sun has eight planets, five dwarf planets and many thousands of minor planets.

Pulsar
A fast-spinning neutron emitting radiation that appears as short pulses if the beam of radiation is in our line of sight. The pulses are extremely regular and range from milliseconds to several seconds. There are more than 1,600 pulsars that have been observed in our Galaxy, the first discovered by Jocelyn Bell in 1967. The Crab Nebula, which is the remnant of the supernova seen in 1054 by the Chinese, has a pulsar that flashes 33 times each second.

Reflector
A telescope that uses a large mirror to collect and focus light.

Refractor
A telescope that uses a large lens to collect and focus light.

Retrograde motion
The outer planets exhibit retrograde

motion as seen from Earth when they move in the sky from east to west rather than the normal west to east motion. This is due to Earth, which has a faster speed than the outer planets, catching up with the outer planet and then pulling away as it orbits the Sun.

Satellite
Any object in orbit around a larger body. Most planets have satellites.

Solar mass
A unit of mass used by astronomers equivalent to the Sun. Stars range in mass from 0.06 of a solar mass to 100 solar masses. The largest galaxies are over a trillion solar masses.

Solar System
Our cosmic backyard, containing the Sun and everything that orbits the Sun, including the Earth, the Moon, the inferior and exterior planets and their satellites, asteroids and comets.

Spectroscope
An instrument that splits up electromagnetic radiation into its different wavelengths and is used to analyse how the intensity of radiation varies with wavelength. With a spectroscope, the astronomer, Sir Norman Lockyer, discovered the element helium in the Sun in 1868 before it was found on Earth.

Spectrum
The splitting of electromagnetic waves into its various component wavelengths, as light is split into the colours of the rainbow, is called a spectrum. A spectrum can be analysed to reveal a great many properties including, among other things, surface temperature, radial velocity, composition and magnetic field strength.

Star
A huge ball of incandescent gas shining by nuclear fusion. The Sun is a star.

Stargazer
An intelligent person who enjoys viewing the night skies, and often contemplates life, the Universe and everything, while doing so.

Sun
Our nearest star.

Sunspot
A slightly cooler region on the surface of the Sun, which appears dark against the brighter background. Most sunspots have a lifetime of less than one rotation of the Sun.

Supernova
The catastrophic explosion of a giant star at the end of its life, which can release as much energy per second as the output of a whole galaxy of 100 billion stars. There have been six supernovae seen with the naked eye in the last 2,000 years.

Telescope
An instrument that collects and focuses electromagnetic radiation – from long wavelength radio waves, through visible light to short-wavelength gamma rays. **Optical telescopes** produce magnified images of distant objects by using lenses and/or mirrors to collect and focus light.

Terminator
The line separating the illuminated and unilluminated hemispheres of a planet or satellite, marking the line of sunrise or sunset.

TLP
Transient lunar phenomena. Rarely observed, short-lived anomalous coloured glows, flashes or obscuration of local surface detail on the Moon whose causes are poorly understood.

Universe
The entire observable cosmos. Everything we know about – the whole shebang.

Variable star
A star whose apparent brightness fluctuates over time, either through being eclipsed by an orbiting companion or through changes in its size and/or the level of its light output.

Zenith
The point in the sky directly above the observer. It has an altitude of 90 degrees.

INDEX

PICTURE CREDITS ....

AAO / Digitized Sky Survey (public domain)
113

Anthony Ayiomamitis
12, 28 (r), 43 (l), 45 (b), 49, 55, 56 (t and b), 63 (l), 66 (b), 68 (b), 129 (r), 130 (t)

Adam Block / NOAO / AURA / NSF (public domain)
88

Mike Brown
131, 132 (t), 143 (b)

Maurice Collins
34 (b), 85, 153

Jamie Cooper
34 (t), 116, 128

ESO (public domain)
93

John Fletcher
42

Akira Fujii (public domain)
92

Peter Grego
8, 10, 11, 14, 16 (b), 17, 20, 21, 22, 24, 25, 26, 32, 33, 38, 40–41, 43 (r), 45 (t), 50–51, 53 (b), 59–59, 60, 62, 63 (r), 64–65, 66 (t), 68 (t), 69, 70 (b), 72–73, 74, 75 (r), 77 (r), 78–79, 85, 86–87, 94–95, 100–101, 105, 107 (b), 108–109, 110, 111, 116, 120, 122, 127, 129 (r), 130 (b), 131, 133 (t), 134, 135, 138, 139, 141 (t), 143 (t), 152

Nick Howes
5, 7, 44, 53 (t), 71, 75 (l), 91, 119 (t), 124, 148, 149

Keck Observatory
146

Simon Kidd
132 (b)

Martin Lewis
141 (b), 145

NASA
2, 19, 21, 22, 23 (l), 23 (r), 29 (l and c), 30 (l and r), 31 (l and r), 81, 122, 123, 133 (b), 136, 137, 140, 142, 144, 147 (t), 151

NASA / STScI
27, 83, 96, 147 (b)

NASA / STScI / Steinberg / Adam Block / NOAO / AURA / NSF
18

Public domain
15, 16 (t), 38 (overlay)

John Rowlands
119 (b)

Siding Spring Observatory (public domain)
89

STScI / DSS (public domain)
90

Alan Tough
118

Peter Vasey
4, 6, 10, 11, 29 (r), 46 (t and b), 47, 48 (t and b), 52, 54, 57, 61, 67, 70 (t), 76, 77 (l), 80, 82, 97, 98, 99, 102, 103, 104, 106, 107 (t), 112, 114, 115, 125, 126, 129 (l)

VLT / ESO
28 (l), 84

Star symbols used throughout Part One:
Denis M Moskowitz and Alec Finlay
http://www.suberic.net/~dmm/astro/constellations.html

A catalogue record for this book is available from the British Library.
ISBN-13: 978-1-4463-0239-2 Paperback
ISBN-10: 1-4463-0239-3 Paperback

ISBN-13: 978-1-4463-5651-7 e-pub
ISBN-10: 1-4463-5651-5 e-pub

ISBN-13: 978-1-4463-5650-0 PDF
ISBN-10: 1-4463-5650-7 PDF

10 9 8 7 6 5 4 3 2 1

Senior Editor: Verity Muir
Project Editor: Ame Verso
Senior Designer: Victoria Marks
Production Manager: Beverley Richardson

Paperback edition printed in China by Toppan Leefung for:
F&W Media International, Ltd
Brunel House, Forde Close, Newton Abbot, TQ12 4PU, UK

F+W Media publishes high quality books on a wide range of subjects.
For more great book ideas visit: www.fwmedia.co.uk